THE ABCs
OF HIGH-PRESSURE
SCIENCE

THE ABCs
OF HIGH-PRESSURE
SCIENCE

Sergei M. Stishov

The P.N. Lebedev Physical Institute
of the Russian Academy of Sciences, Russia

 **World Scientific**

NEW JERSEY · LONDON · SINGAPORE · BEIJING · SHANGHAI · HONG KONG · TAIPEI · CHENNAI · TOKYO

Published by

World Scientific Publishing Europe Ltd.

57 Shelton Street, Covent Garden, London WC2H 9HE

Head office: 5 Toh Tuck Link, Singapore 596224

USA office: 27 Warren Street, Suite 401-402, Hackensack, NJ 07601

Library of Congress Cataloging-in-Publication Data

Names: Stishov, Sergei M., author.

Title: The ABCs of high-pressure science / Sergei M. Stishov, P.N. Lebedev Physical Institute
of Russian Academy of Sciences, Russia.

Description: New Jersey : World Scientific, [2021] | Includes bibliographical references and index.

Identifiers: LCCN 2020041566 (print) | LCCN 2020041567 (ebook) |
 ISBN 9781786349552 (hardcover) | ISBN 9781786349569 (ebook) |
 ISBN 9781786349576 (ebook other)

Subjects: LCSH: High pressure (Science)

Classification: LCC QC280 .S75 2021 (print) | LCC QC280 (ebook) | DDC 531/.1--dc23

LC record available at https://lccn.loc.gov/2020041566

LC ebook record available at https://lccn.loc.gov/2020041567

British Library Cataloguing-in-Publication Data

A catalogue record for this book is available from the British Library.

For any available supplementary material, please visit
https://www.worldscientific.com/worldscibooks/10.1142/Q0281#t=suppl

Desk Editors: George Vasu/Michael Beale/Shi Ying Koe

Typeset by Stallion Press
Email: enquiries@stallionpress.com

Preface

Any professional community develops its own jargon to describe its terms and concepts. This always makes it difficult for a novice to enter the profession. This short book is for students, graduate students, scientists and engineers new to high-pressure science and technology. It provides brief information on most of the topics and concepts encountered in high-pressure research and engineering. Literary references are provided where it is absolutely necessary. Interested readers can obtain more information on anything else from the Internet. However, I would like to recommend three useful books devoted to high-pressure techniques:

1. *Handbook of Techniques in High-Pressure Research and Engineering* by Tsiklis (1968);
2. *High Pressure Technology*, Vols. 1 and 2, by Spain and Paauwe (1977);
3. *Experimental Techniques in High-Pressure Research* by Sherman and Stadtmuller (1987).

About the Author

Sergei M. Stishov is a physicist-experimentalist, born in 1937. He received his academic training from Moscow State University, getting his PhD in Geochemistry in 1960. He became a Doctor of Science in Physics in 1974. He spent his early years at the Institute of Crystallography, Moscow (1962–1993), and later he became the Director of the Institute for High Pressure Physics, Moscow (1993–2016). Currently, he is a staff member of the Physical Lebedev Institute  (Moscow). Stishov has done research on phase transitions at high pressures, high Tc superconductivity, and strongly correlated electronic systems. His most famous work was a discovery of the dense silica (1961), the natural analog of which was named "stishovite". He has been awarded the P.W. Bridgman Gold Medal (2005) and the Petr Kapitza Gold medal (2014).

Contents

F 47

G 49

H 55

I 69

A

AIRAPT

AIRAPT — International Association for the Advancement of High-Pressure Science and Technology (Association Internationale pour l'Avancement de la Recherche et de la Technologie aux Hautes Pressions). The first AIPART conference took place in Le Creusot (France) in 1965 through the efforts of Boris Vodar, a French scientist with Russian roots. Since then, the AIRAPT Conferences have been held every two years. At the conference, two awards are presented: the Bridgman Gold Medal (**Bridgman Gold Medal**) and the Jamieson Award for young scientists (**Jamieson Award**).

Akimotoite

Akimotoite is a mineral of composition $(Mg, Fe)SiO_3$ with a crystal structure of ilmenite, named after a Japanese high-pressure researcher, Sun Ichi Akimoto (1925–2004). It contains silicon in the six-fold coordination. Found in the meteorite Tenham (Australia). Akimotoite, along with Bridgmanite (**Bridgmanite**), is the high-pressure phase of the pyroxene composition.

Amagat

Emile Hilaire Amagat (1841–1915) was a French physicist, member of the Paris Academy of Sciences, known for his experimental studies on the equations of state of noble gases at pressures up to 3,000 atm. and 200°C. Along with this, he was engaged in the development of high-pressure equipment. Apparently, his most well-known invention is the method of making electrical lead-throughs using an inverted cone (Amagat cone)[1] (see **Electrical Leads**). The **Amagat** is a unit of number density: the number of Ideal Gas molecules per unit volume at 1 atmosphere and 0°C.

Amorphization at High-Pressure

Amorphization at high-pressures often occurs when substances with an open-crystal structure are subjected to pressure at a relatively low temperature. In this case, the phase transition to the denser phase does not occur owing to kinetic reasons. The high stresses, figuratively speaking, simply break the structure, creating an amorphous state. Examples are the amorphization of ice, quartz and graphite (see **Graphite**).

Amorphous Carbon

Amorphous carbons are substances consisting mainly of carbon materials not having a crystalline structure. These materials contain a variable amount of sp, sp^2 and sp^3 hybridized bonds, thus exhibiting graphite-like or diamond-like properties. Soot, natural shungites, deposited carbon films and superhard carbon coatings are examples of such materials.

Andrews

Thomas Andrews (1813–1885), an Irish physical chemist, member of the Royal Society of London, discovered the phenomenon of continuous transition between the gaseous and liquid states of matter. Andrews's research served as a basis for the creation of the van der Waals theory of critical phenomena (see **Critical Point**).

---

[1]Bridgman P. W., *The Physics of High-Pressure*. London, G. Bell and Sons, Ltd. (1931).

Anti-extrusion Ring

Anti-extrusion ring is a metal ring that prevents the extrusion of a soft gasket (see **Seal**).

Argon

Argon (Ar) is a noble gas with a critical temperature 150.9 K. It becomes liquid at 87.3 K and then solid at 83.8 K. At 25°C, fluid argon crystallizes at pressure of ~15 kbar, which makes argon an important medium for creating hydrostatic environment. Under pressure, argon can form compounds of the type $Ar(H_2)_2$, $Ar(O_2)_3$.

Atmosphere

Atmosphere is a unit of measurement of pressure. There are physical and technical atmosphere: 1 phys. atm = 760 mm Hg = 1.0332 tech. atm or kg/cm^2 = 1.0133 bar (**Pressure — Definition and Units**).

Atomic Bomb

There are two types of atomic bombs based on the fission chain reaction: uranium (gun-type) and plutonium (implosion-type). In the first of these, the critical mass is achieved by firing a uranium projectile (^{235}U) into a uranium (^{235}U) target, thereby forming a critical mass. In the second case, the supercritical volume of plutonium in the form of a hollow sphere turns to a critical state as a result of compression by strong shock waves arising from the explosion of a shell of detonating material. In principle, a uranium bomb can also be detonated by implosion, but not vice versa. The explosion of a plutonium bomb cannot be accomplished by shooting at a target, since the rate of spontaneous fission in plutonium is too high. The latter makes it impossible to achieve a sufficient speed of a Pu-projectile to a Pu-target to prevent a thermal explosion. Note that the study of strong shock waves started in 1940s nowadays goes far beyond its military applications (**Dynamic Pressure**).

Autoclave

It is impossible to heat water above 100°C at normal pressure. However, if the water temperature needs to be raised — increasing the solubility of substances or for sterilization — then the heating of water or solution is carried out in a confined space. In this case, the pressure of the medium increases and in order to keep it within a given volume, special devices — autoclaves — capable of withstanding the arising pressure, are used. The range of these devices is very wide and extends from pan's pressure cookers and medical sterilizers to huge autoclaves for growing quartz crystals (**Hydrothermal Process**). High-pressure autoclaves designed for crystal growing, studying chemical and mineral equilibria are usually thick-walled vessels made of heat-resistant steels with self-sealing closures based on the principle of unsupported area (see **Seal**). Autoclaves in these applications typically operate at several thousand atmospheres and hundreds of degrees Celsius.

Autofrettage

Autofrettage is a procedure for strengthening cylindrical thick-walled vessels, which involves plastic deformation of the inner layers of vessels by means of pressure exceeding the yield pressure. As a result, residual compressive stresses occur in the outer layers of the vessel permitting a higher operating pressure.

B

Bar

Bar — unit of pressure, 1 bar $= 10^6$ dyne/cm^2 (**Pressure — Definition and Units**).

Belt

The Belt is a high-pressure apparatus invented by Tracy Hall in 1954 (see **Hall**). It was with this apparatus that the first artificial diamonds were obtained in the laboratory of the General Electric Company. The device consists of two conical pistons and a matrix containing the heater and the test substance or special mixture (Fig. 1). This matrix has a specific form, called Belt by the inventor. All the most loaded parts of the apparatus are made of carboloy (tungsten carbide cemented with cobalt). The Belt was classified by the US Government and the GE company, and it was only in 1960 that T. Hall was able to describe the device in detail.[1]

Bernal

John Despond Bernal — a British crystallographer (Fig. 2) who graduated from Cambridge University in 1922. Until 1927, he conducted research under the direction of William Bragg in London.

---

[1]Hall H. T., *Rev. Sci. Inst.*, 31, 125 (1960).

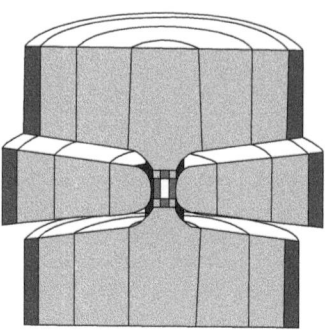

Fig. 1.   High-pressure apparatus "Belt".[2]

Fig. 2.   John Bernal (1901–1971).

Returning to Cambridge in 1927, Bernal initiated a program to study complex biological molecules, which eventually led to the Nobel level studies, carried out by his students and collaborators, among which Rosalind Franklin, Dorothy Hodgkin, Aaron Klug and Max Perutz should be mentioned. However, in the present context, two of his ideas with far-reaching consequences for research on high-pressures should be noted. First, Bernal turned the attention of geophysicists to the possible polymorphism of olivine Mg_2SiO_4,[3,4] which laid the foundation for experimental studies of phase transitions in the depths of the Earth. Then he expressed the idea that all substances transform into a metallic state at high pressures.[5] This idea inspired Wigner and Huntington to conduct the first calculations of metal transition in hydrogen.[6]

[2]Hall H. T. High pressure apparatus, in *Progress in Very High Pressure Research*, edited by F. P. Bundy, W. R. Hibbard and H. M. Strong, New York: John Wiley and Sons, Inc. (1961).
[3]Bernal J. D. *Observatory*, LIX, 268 (1936).
[4]Ringwood A. E. *Geochem. et Cosmochim. Acta*, 13, 303 (1958).
[5]Wigner E. and Huntington H. B., *J. Chem. Phys.*, 3, 764 (1935).
[6]*Ibid.*

Beryllium Bronze

Beryllium bronze — an alloy of copper and beryllium ($\sim$2.5% Be). It has high strength after heat treatment (quenching into water from 780°C, annealing at 320°C). Its hardness after heat treatment is HRc $\sim$38–40. Beryllium bronze is used for the manufacture of non-magnetic high-pressure vessels, vessels for working with hydrogen, anti-extrusion and anti-friction rings.

Birch

Albert Francis Birch — an American geophysicist (Fig. 3), graduated from Harvard University in 1924, receiving a Bachelor's degree in electrical engineering. After working for some time in the industry, he went to Strasbourg, where he continued his studies under the general guidance of Pierre Weiss. In 1926, Birch returned to Harvard, where he received a Master's degree in 1928, this time in the field of physics and entered graduate school. In 1932, Birch defended his thesis for the degree of Doctor of Philosophy (PhD) under the leadership of Percy Bridgman, the future Nobel Prize winner. Birch's

Fig. 3. Francis Birch (1903–1992) (Harvard University archive).

dissertation work was connected with the study of the critical point on the boiling curve of mercury. However, the subsequent work of Birch was related to the problems of the Earth's structure. In 1947, Birch, based on the Murnaghan theory of finite deformations, proposed an equation of state, called the Birch–Murnaghan equation.[7] The analysis of this equation, in combination with the experimental data obtained at low pressures, allowed Birch to determine the nature of the behavior of the elastic properties of matter at pressures corresponding to the depths of the Earth. In 1952, Birch published his most important work on the composition and elastic properties of the Earth's depths.[8] In this work,

---

[7]Francis Birch, *Phys. Rev.*, 71, 809 (1947).
[8]Francis Birch, *J. Geophys. Res.*, 57, 227 (1952).

Birch showed that the Earth's mantle is not homogeneous, and an abnormal change in the elastic properties of the mantle, occurring at depths of the order of 400 km, could not be explained either by the natural compression of the Earth's matter or by the change in its chemical composition. The fact is that the ratio of the bulk modulus to the density K/ρ in the region of depths of 400 km and more turned out to be so high that it could not result from the compression of minerals and rocks available on the surface of the Earth. In order to explain this observation, Birch put forward a hypothesis about phase transformations in the Earth's interior caused, in particular, by the transition of silicon to the sixfold coordination with respect to oxygen in silicate substances. That would ensure a proper high ratio K/ρ. This hypothesis was brilliantly confirmed.

In wartime, Birch, as a high-pressures expert, was one of the main developers of the atomic bomb of the "gun" type. In this type of a bomb, a corresponding gun fired a uranium blank at a uranium target, thereby creating a supercritical mass. To create the bomb dropped on Hiroshima (Little Boy), it was necessary to use all the available amount of ^{235}U (64.1 kg).[9]

Booster

Booster — can be used as a name for a pressure intensifier (see **Intensifier**).

Borazon (BN)

Borazon is a cubic form of boron nitride with a crystal structure of the zinc blende (ZnS). It was made for the first time by Robert Wentorf in 1957.[10] Borazon is a superhard material, widely used in industry for the processing of steel (does not interact with iron, unlike diamond). In his publication,[11] Wentorf did not disclose all details of the borazon synthesis, which led the followers to a great disappointment when trying to synthesize borazon using 8-group metals

---

[9]Ahrens, Thomas J. Albert, *Francis Birch. Biographical Memoirs.* 74. Washington, D.C.: National Academy of Sciences (1998).
[10]Wentorf R. H., *J. Chem. Phys.*, 26, 956 (1957).
[11]*Ibid.*

as catalysts. In fact, alkali and alkaline Earth metals are used in the borazon synthesis.

Boron (B)

Boron is a light element (atomic number 5) of the third group of the second period of the Mendeleev periodic system. It is commercially available as fine crystals and amorphous powder. Due to the relative transparency for X-rays, it is used (sometimes mixed with epoxy) for manufacture of windows for X-ray structural studies.

Boron Nitride (BN)

Boron nitride (BN) is a compact, inert and heat-resistant material of white color with a graphite-like crystalline structure. It is easily machinable and used as a chemically inert, heat-resistant insulator in high-pressure, high-temperature apparatuses.

Bourdon Manometer

The bourdon manometer is a secondary manometer, which works due to the elastic deformation of the measuring tube (Bourdon tube) under the action of pressure. The Bourdon tube is a flattened, curved tube that tends to straighten up under internal pressure (Fig. 4).

Boyle

Robert Boyle (1627–1661) was an English physicist, chemist and theologian. He established the law of gas compression in the form $pV = \text{const}$, known as the Boyle–Mariotte law.

Bridgman

Percy Williams Bridgman, an American physicist and philosopher, Professor of Harvard University (Figs. 5 and 6), was awarded the Nobel Prize in 1946 with the citation "for the invention of an

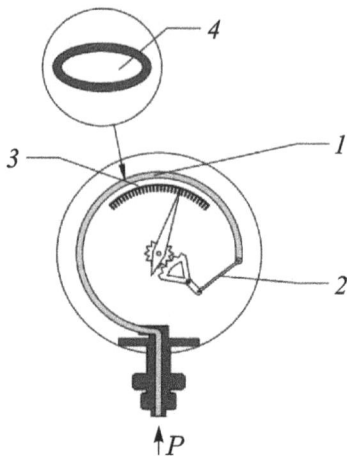

Fig. 4. Bourdon manometer: 1 — Bourdon tube, 2 — leash, 3 — scale, 4 — sectional view of the Bourdon tube.

Fig. 5. P. W. Bridgman (1882–1961) (Harvard University archive).

apparatus to produce extremely high-pressures and for the discoveries he made therewith in the field of high-pressure physics". Bridgman graduated from Harvard University (Cambridge, Massachusetts, USA) in 1904. He then began teaching at Harvard. In 1908, he received his PhD. In 1919, he became a full professor. Bridgman's students were such significant figures as Francis Birch, Robert Oppenheimer and John Slater.

Since 1905, Bridgman explored the properties of substances at high pressures. The unreliability or lack of appropriate equipment forced him to engage in the development of new approaches to high-pressure experimental technology. These included the principle of unsupported area, which he used as the basis for movable and stationary seals, the method of variable mechanical support of high-pressure vessels, the principle of massive support implemented in the so-called Bridgman anvils, and much more. With the help of advanced technology, Bridgman conducted systematic studies of the physical properties of many

Fig. 6. P. W. Bridgman at work on the "thirty-thousandth" installation, which allowed him to regularly receive 30,000 atmospheres (Harvard University archive).

substances and compounds up to pressures of 100 kbar. These works of Bridgman are partially summarized in his book *Physics of High Pressure*.[12] One of the methods of growth of single crystals developed by Bridgman bears his name. The complete collection of the experimental work of Bridgman was published in seven volumes by Harvard University in 1964.[13] The name of Bridgman is also carried by the gold medal awarded by the International Association for the Advancement of High-Pressure Science and Technology (AIRAPT).

P. W. Bridgman is also the author of several books on the general issues of physics and philosophy listed below:

(1) *Dimensional Analysis*, New Haven, Yale University Press, 1922.
(2) *A Condensed Collection of Thermodynamic Formulas*, Cambridge, Harvard University Press, 1925.

---

[12]Bridgman P. W., *The Physics of High Pressure*. London: G. Bell and Sons, Ltd. (1931).
[13]Bridgman P. W., *Collected Experimental Papers*. Cambridge, Massachusetts: Harvard University Press (1964).

(3) *The Logic of Modern Physics*, New York, The Macmillan Company, 1927.
(4) *The Thermodynamics of Electrical Phenomena in Metals*, New York, The Macmillan Company, 1934.
(5) *The Nature of Physical Theory*, Princeton University Press, 1936.
(6) *The Intelligent Individual and Society*, New York, The Macmillan Company, 1938.
(7) *The Nature of Thermodynamics*, Cambridge, Harvard University, 1941.
(8) *Reflections of a Physicist*, New York, Philosophical Library, Inc., 1950.
(9) *The Nature of Some of Our Physical Concepts*, New York, Philosophical Library, 1952.
(10) *Studies in Large Plastic Flow and Fracture, with Special Emphasis on the Effects of Hydrostatic Pressure*, New York, McGrow-Hill Book Co., 1952.
(11) *The Way Things Are*, Cambridge, Harvard University Press, 1959.
(12) *A Sophisticate's Primer of Relativity*, Middletown, Connecticut, Wesleyan University Press, 1962.

Bridgman Anvils, Lentils, Toroid

Bridgman anvils, lentils, toroid are systems of two truncated conical pistons directed towards each other (opposed anvils). Figure 7(a) shows the construction proposed by Bridgman in 1935. The so-called principle of massive support is implemented in this system, when the contact area of two anvils, which experience the greatest stresses, is supported by massive parts of the apparatus. The proposed system with pistons of carboloy withstands a contact pressure of ~200 kbar, while the tensile strength of the corresponding cylindrical pistons is about 60 kbar. It is obvious that the use of Bridgman's anvils is associated with certain limitations, for example, due to the difficulties of using an internal heater. It is for this reason P. Bridgman could not synthesize diamonds.

However, during operations, it turned out that under heavy loads, the Bridgman anvils plastically deform in such a way that the flat contact surface turns into a concave, almost spherical segment. These

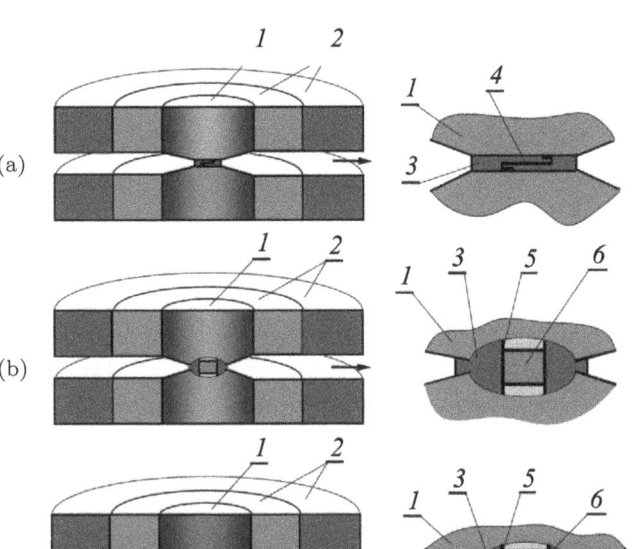

Fig. 7. Bridgman's anvils (a), lentils (b), toroid (c): 1 — conical pistons made of hard alloy or tool steel, 2 — supporting steel rings, 3 — pressure transmitting medium (pyrophyllite, lithograph stone), 4 — wire sample, 5 — short-circuited heater, 6 — sample, 7 — electrical lead-throughs.

deformed anvils put together form a quite noticeable cavity. Usually, deformed anvils were polished and returned to their original state, but at the Institute of High-Pressure Physics of the Russian Academy of Sciences (HPPI), it was decided to turn this problem into a blessing. At the Institute, profiled anvils were made, forming a cavity sufficient for introducing a short-circuited graphite heater. The shape of the cavity resembled a lentil grain, which determined the name of this cell (Fig. 7(b)). Subsequently, the cell design was supplemented with a toroidal cavity, filled with the pressure transmitting medium. The new device was called the "toroid" (Fig. 7(c)). The principle of massive support is widely used in the design of high-pressure apparatuses (see **Belt, Multi-anvil Installations**).

Bridgman Gold Medal

The Bridgman Gold Medal (see **Bridgman**) was established by the International Association for the Advancement of High-Pressure

Science and Technology (AIRAPT). For the first time, in 1977, the Bridgman medal was awarded to H. Drickamer (see **Drickamer**) at the **AIRAPT** Conference in Boulder (USA).

Bridgmanite

Bridgmanite, a mineral of the composition $(M, Fe)SiO_3$ with a crystal structure of perovskite, is a high-pressure phase of ordinary pyroxene. It was found in the Tenham meteorite (Australia) and named after the high-pressure pioneer Percy Bridgman (see **Bridgman**). It is believed that this mineral constitutes about 90% of the lower mantle.

C

Calibration

Calibration is the procedure for assigning an arbitrary measurement scale of any value to an absolute value or an absolute scale (for example, calibrating the electrical resistance of a manganin gauge using a deadweight pressure gauge (see **Deadweight Pressure Gauge**)).

Carbin

Carbin is a linearly polymerized form of carbon.

Chao

Edward Ching-Te Chao — an American geologist and mineralogist (Fig. 8) — received a PhD in geology from the University of Chicago in 1948. In 1949, he began working in the US Geological Survey, where he worked until his retirement in 1994. Chao discovered two dense forms of silica: coesite[1] and stishovite,[2] in the Arizona impact crater and later in the Reese crater in Germany. This discovery marked the emergence of a new branch of mineralogy — high-pressure mineralogy. The minerals found by Chao serve

---

[1]Chao E. C. T., Shoemaker E. M. and Madsen B. M., *Science*, 132, 220 (1960).
[2]Chao E. C. T., Fahey J. J., Littler J. and Milton D. J., *J. Geophys. Res.*, 67, 419 (1962).

Fig. 8.   Edward Chao (1919–2008) (from S. M. Stishov archive).

as indicators of the impact origin of various geological structures. The asteroid "3906 Chao" and the mineral "chaoite" of impact origin, found in the crater of Reese (Germany), are named after Edward Chao.

Check Valve

Check valve — a device that allows liquid or gas to flow only in one direction — is an integral part of pumps and compressors. In Fig. 9, there are several types of valves used in the laboratory: the simplest ball valve (a), the cone valve (b), the poppet valve (c), the poppet valve with the rubber ring (d). As P. Bridgman noted, the ball valve wears out quickly, acquiring an annular groove in the place of contact, and turned, ceases to work reliably. The cone valve works well in liquid pumps up to several kilobars. The poppet valves can work with gas pressures, but they are very sensitive to even small particulate contamination. The valve (d) is devoid of this disadvantage, it works reliably under conditions of high gas pressure until elasticity of the sealing ring is lost. The presented valves can operate in the mode of both suction and discharge, or be used as safety valves depending on the geometry of the location, spring force and operating pressure.

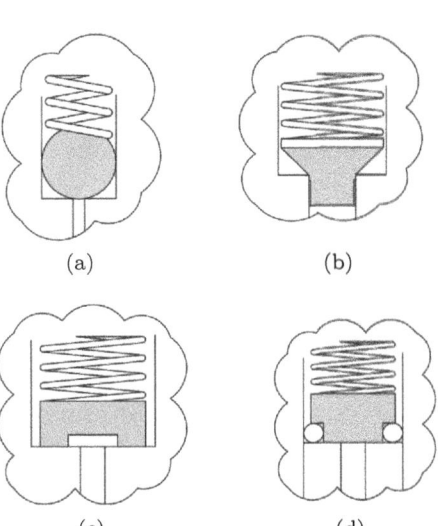

Fig. 9. Check valves: (a) ball valve, (b) conical valve, (c) poppet valve, (d) poppet valve with O-ring (see **O-ring**).

Chemical Transformations

Without regard to polymerization and synthesis reactions at relatively low pressures, regularly carried out in the chemical industry, I will point out several examples of chemical transformations occurring at high pressures. We note here that at very high pressures the contribution of the term $P\Delta V$ (here P pressure, V is volume) to the thermodynamic equilibrium condition becomes decisive. For this reason, by subjecting NaCl to high pressures and temperatures in the presence of an excess of Cl and an excess of Na, researchers obtained unusual compounds $NaCl_3$ and Na_3Cl, which do not correspond to the classical concepts of valence.[3] Strange hydride H_3S is formed by the decomposition of hydrogen sulfide H_2S at high pressures.[4] An example of geophysical significance associated with the decomposition of olivine at high pressures

---

[3]Zang W. *et al.*, *Science*, 342, 1502 (2013).
[4]Goncharov A. F. *et al.*, *Phys. Rev. B*, 93, 174105 (2016).

according to the reaction[5] is (Mg, Fe)$_2$SiO$_4$ = (Mg, Fe)SiO$_3$ + + (Mg, Fe)O. In this case, a silicate phase with a perovskite structure is formed, containing silicon in the six-fold coordination, which is quite different from silicates of surface minerals and rocks.

Coesite

Coesite is a dense modification of silica (the density of coesite is 3.01 g/cm^3, quartz is 2.65 g/cm^3), first obtained by L. Coes in the laboratory. Subsequently, in 1960, Edward Chao (see **Chao**) discovered coesite in the Arizona meteorite crater (see **Crater**). The crystal structure of coesite is not fundamentally different from the structures of all other modifications of silica known at atmospheric pressure, and is a network of silicon — oxygen tetrahedra. Coesite is found in almost all craters of impact origin and in a number of metamorphic rocks that have risen from great depths.

Compressor

A compressor is a high-speed machine of continuous action designed to compress gases. It usually uses a large amount of liquid lubricant. As a result, the compressible gas is heavily polluted. Bulky filters are built to clean the gas, which do not always save the situation. The next section describes the membrane compressor, which is a source of pure compressed gas.

Compressor Membranic

Figure 10 shows a membrane compressor, the design of which is based on application of a flexible steel membrane (4). The compressor works the following way. When the piston (2) moves "down", the oil is sucked into the compressor through the valve (1), while the cavity located above the diaphragm is filled with gas through

---

[5]Wang Y. *et al.*, *Science*, 275, 510 (1997).

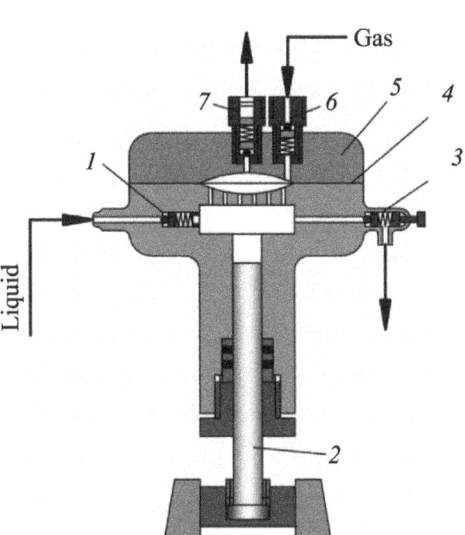

Fig. 10. Schematic view of a membrane compressor: 1, 3 — liquid suction and control valves, 2 — piston, 4 — membrane, 5 — steel block, 6, 7 — suction and injection gas valves.

the valve (6). When the piston (2) moves "up", the oil pushes the diaphragm, which injects gas through the valve (7) into the appropriate container. Excessive oil pressure is released through the control valve (3). In fact, the whole process can be described as a rapid pulsation of the membrane, which actuates the valves (6) and (7), and ultimately compresses the gas to high pressures determined by the strength of the membrane and the properties of the pump.

Crater

A crater is a terrain form that appears as a result of volcanic events or impact of a large meteorite or an asteroid (see Fig. 11).

The impact of a meteorite is accompanied by the raising of pressures and temperatures, which leads to the birth of minerals typical of high-pressure and -temperature conditions. These include: diamond, coesite, stishovite (see **Diamond, Coesite, Stishovite**). A huge number of craters of impact origin are found on the Earth's

Fig. 11.   Meteorite crater in the Arizona desert.

surface. Some of them have diameters of the order of hundreds of kilometers. The most famous of them is the Chicxulub crater in Mexico, about 180 km in diameter, which is associated with the extinction of dinosaurs. The crater Popigai in Siberia with a diameter of about 100 km contains billions of carats of polycrystalline diamonds that arose when a meteorite hit a carbon-rich rock.

Critical Point

The critical point is a multi-valued name used in various fields of science and technology. In the field of high pressures, of interest is

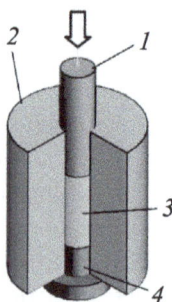

Fig. 12.   Cylinder–piston system: 1 — piston, 2 — cylinder, 3 — substance under study, 4 — plug.

the critical point of a second-order phase transition or a critical point of the type: liquid – vapor, in a first-order phase transition, where the difference between the two phases disappears.[6]

Cylinder–Piston

Cylinder–piston — a term denoting the type of high-pressure system, shown in Fig. 12.

---

[6]Landau L. D. and Lifshitz E. M., *Statistical Physics*, Vol. 5 (3rd edn.). Butterworth-Heinemann (1980).

Daphne Oil

Daphne oil — industrial oil with an exotic name (Daphne — the name of a naiad from the Greek mythology), is used as a medium transmitting pressure, especially in experiments at low temperatures.

De Broglie's Thermal Wavelength

De Broglie's thermal length $\lambda_T = \sqrt{\frac{2\pi\hbar^2}{mkT}}$ defines the wave properties of the particles as a function of temperature. Here $\hbar = h/2\pi$, h is the Planck constant, T is temperature, m is mass.

Deadweight Pressure Gauge

The deadweight pressure gauge is a device for absolute pressure measurements. It also serves to calibrate secondary pressure sensors: Bourdon pressure gauges (see **Bourdon Manometer**), Manganin pressure gauges (see **Manganin Gauge**), capacitance and strain gauge pressure sensors. The principle of operation of the deadweight pressure gauge is to balance the liquid column under pressure with a calibrated load, as shown in Fig. 13. As can be seen in the figure, the weights (3) rest on the piston, which moves freely in the high-pressure vessel. The gap between the piston (1) and the walls of the vessel (2) is made very small, so with a sufficient viscosity of the fluid, the leakage is not a problem. When equilibrium is reached, the

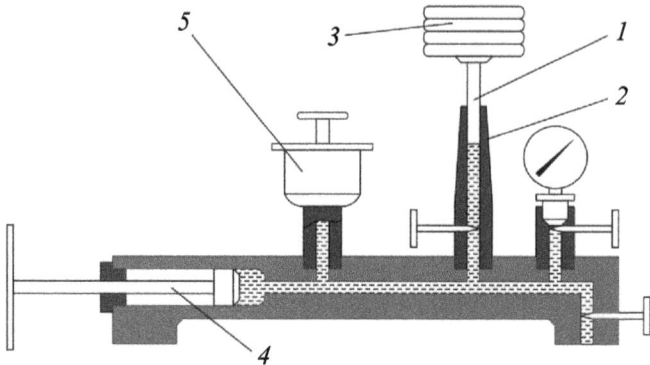

Fig. 13.   General view of a deadweight pressure gauge: 1, 2 — piston pair, 3 — calibrated load, 4 — screw press, 5 — oil tank, the rest are valves and pressure gauge.

piston with weights is only very slowly lowered due to small leaks. The measured pressure is calculated using the elementary formula $P = F/S_{\text{eff}}$, where F is the weight of the load and the piston, S_{eff} is the effective area of the piston, calculated taking into account all possible corrections, such as: a layer of liquid adjacent to the piston, deformation of the vessel and the piston, etc. To measure high pressures of the order of several kilobars, a differential piston system is used to limit the weight of the calibration load to reasonable values. The limit of measured pressures with a dead weight pressure gauge does not seem to exceed 20 kbar.

Deformation

Deformation is a change of shape or size of a body under an influence of external forces (stresses) and temperatures. There are elastic and plastic deformations. In the first case, the deformation is reversible, while in the second, there is a residual deformation. Mechanical properties of metallic materials are usually characterized using stress–strain diagrams (see Fig. 14). At low stresses, the deformations are linear in stresses (Hooke's law). The point, corresponding to the deviation from Hooke's law is called the limit of proportionality. The point where plastic deformation occurs is called the yield point. In some cases, the material is hardened as a result of plastic

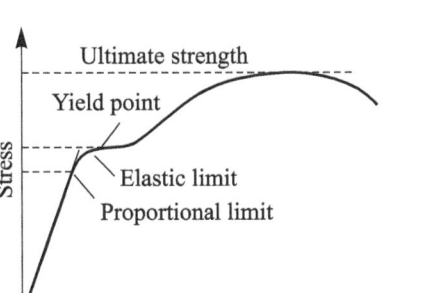

Fig. 14. The generalized stress–strain diagram of a sample of structural steel: A decrease of strength for large strains is associated with the formation of a "neck" at rupture and a decrease in the real diameter of the sample.

deformation. However, in any case, upon further loading, a rupture occurs at a point called the tensile or ultimate strength.

The diagram shown in Fig. 14 characterizes the structural steel, Cu–Be, Ti and Al alloys. Tool steels, hard alloys (WC + Co), diamond compacts, etc., possess high compressive strength, but break at moderate stretching.

Density Functional Theory (DFT)

The DFT is a numerical method for solving many-electron problems. It is widely used in calculations in condensed matter physics.[1] There are a number of relevant computer programs available.

Deuterium (D_2)

Deuterium is a heavy isotope of hydrogen. The deuterium nucleus contains a proton and a neutron. Due to the large difference in the masses of hydrogen and deuterium, the properties of substances consisting of or containing a large amount of hydrogen and deuterium can vary significantly (see **Isotopic Effects**, Fig. 55).

---

[1]Walter Kohn, Nobel Lecture, January 28, 1999, Electronic Structure of Matter — Wave Functions and Density Functionals (https://www.nobelprize.org/prizes/chemistry/1998/kohn/lecture/).

Diamond

Diamond is a transparent modification of carbon with unique hardness and thermal conductivity. Diamonds are widely used in industry for stone cutting, grinding, polishing, drilling, turning and wire drawing, which is not a complete list of the uses of diamond. In recent years, diamond single crystals have been used as the main element of diamond anvils for creating very high pressures. It should also indicate the potential use of nanodiamonds (see **Nanodiamonds**). Natural diamond is found in limited quantities on almost all continents in the kimberlite rocks, rising from great depths, forming the so-called kimberlite pipes. One of these pipes is shown in Fig. 15.

The lack of diamonds for industrial applications set the task of their artificial production. Thermodynamic measurements and calculations[2–5] (see Fig. 16) made it possible to determine the region of diamond stability corresponding to high pressures. Thus, the task of obtaining diamonds in the laboratory required the creation of the necessary equipment.

Fig. 15.   Kimberlite pipe "Mir" Siberia, Russia.

[2]Rossini F. D. and Jessup R. S., *J. Res. Natl. Bur. Stand.*, 21, 491 (1938).
[3]Leipunsky O. I., *Chem. Adv.*, 8, 1519, 1939 (in Russian).
[4]Liljeblad R., *Arkiv for Kemi*, 8, 423 (1955).
[5]Berman R. and Simon F., *Z. Elektrochem.*, 59, 333 (1955).

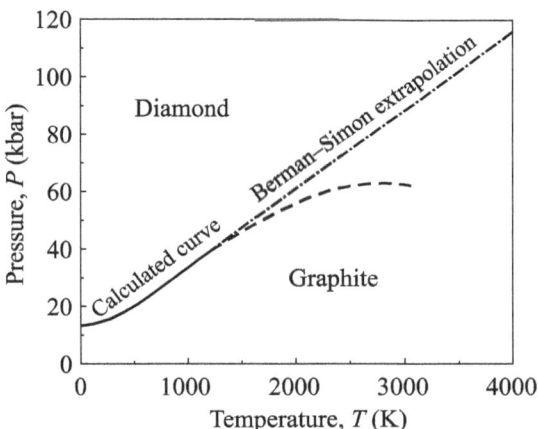

Fig. 16. Graphite-diamond equilibrium line. Dashed line-extrapolation of the Swedish diamond-hunting team.

Source: Bundy F. P., Bovenkerk H. P. *et al.*, *J. Chem. Phys.*, 35, 383 (1961).

This problem was solved, and artificial diamonds were created almost simultaneously in the laboratories of the General Electric company (USA)[6] and the ACEA company (Sweden)[7] in the fifties of the last century. The growth of diamond single crystals, apparently for the first time, was carried out at General Electric[8] (Fig. 17).

Diamonds may play a significant role in the formation of the interior of Neptune, Uranus and exoplanets. The fact is that, despite the openness of the crystal structure (Fig. 18), diamond, due to the small ionic radius of carbon (0.15 Å), has the highest atomic density $(1.73 \times 10^{23}\ cm^{-3})$ among all available and imaginable substances. It is this circumstance that determines its surprising stability at high pressures, as follows from the phase diagram of carbon constructed in[9] and supported by shock-wave experiments (Fig. 19).[10]

---

[6]Bundy F. P., Hall H. T., Strong H. M. and Wentorf R. H., *J. Nature*, 176, 51 (1955).

[7]Liander H., *ASEA J.*, 28, 97 (1955).

[8]Strong H. M. and Wentorf R. H., Jr., *Naturwissenschaften*, 59, 1 (1972).

[9]Correa A. A. *et al.*, *PNAS*, 103, 1204 (2006).

[10]Eggert J. H. *et al.*, *Nat. Phys.*, 6, 40 (2010).

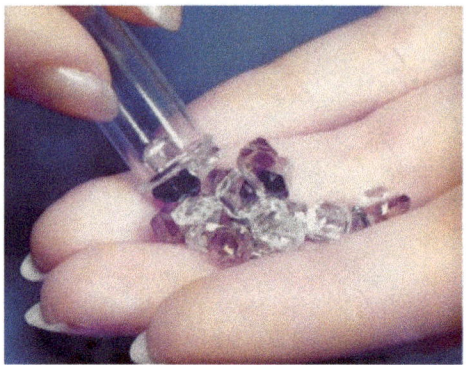

Fig. 17.   Diamond single crystals grown in the laboratory of the General Electric company (archive of the General Electric laboratory).

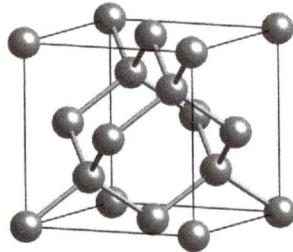

Fig. 18.   Diamond crystal structure model.

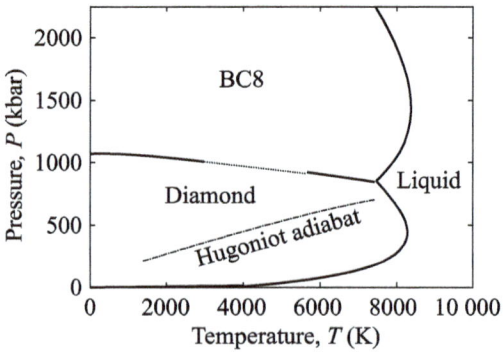

Fig. 19.   Phase diagram of carbon at extreme pressures.
Source: Correa A. A. *et al.*, *PNAS*, 103, 1204 (2006).

Diamond Anvils

The diamond anvil cell is a device for creating very high pressures using diamonds to generate a stress state (pressure) in a sample. The principle of action of the diamond anvil cell is as follows. Two diamonds of gem quality in the form of truncated cones (1) are compressed with the help of various mechanical devices, as shown in Fig. 20. A metal gasket (2) (**Gasket**) is placed between the diamonds with a hole to accommodate a sample (3) and a pressure sensor (4).

As a rule, the hole is filled with a medium (liquid or compressed gas) that transmits pressure. The diameter of the flat working surface of the anvils can vary from fractions of a millimeter to several microns depending on the magnitude of the pressures obtained. The anvil shape can be specially profiled in order to achieve maximum pressures. The material of the gasket can be stainless steel, beryllium bronze, hard steel, rhenium.

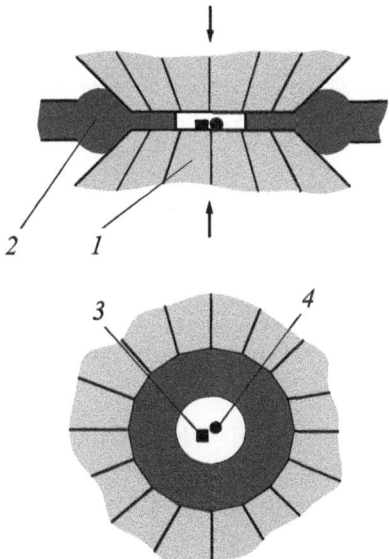

Fig. 20. Diagram of the action of diamond anvils: 1 — diamond, 2 — metal gasket, 3 — sample, 4 — pressure sensor.

Fig. 21.   Charles Weir (left) and Van Valkenburg (right) — inventors of diamond anvils (archive of the National Institute of Standards and Technology (NIST)).

Fig. 22.   First diamond anvils made at the National Bureau of Standards, USA (archive of the National Institute of Standards and Technology (NIST)).

Diamond anvil cells were developed at the National Bureau of Standards (now National Institute of Standards and Technology, NIST) of the USA in 1959 by C. Weir, A. Van Valkenburg and others[11] (see Fig. 21). Figure 22 shows the first version of the diamond anvil device made by Weir hands. This cell is currently on

---

[11]Weir C. E., Lippincott E. R., Van Valkenburg A. and Bunting E. N., *J. Res. Natl. Bur. Stand (US)* 63A, 55 (1959).

display at the NIST Museum. It is noteworthy that Weir and Valkenburg, as employees of a government agency, had access to confiscated
contraband diamonds, which made their task much easier.

Initially, diamond anvils seemed to be a tool with very limited
capabilities. The decisive step was taken by Van Valkenburg, who
proposed using a gasket between the anvils, which allowed a sample
to be placed in the gasket hole filled with the pressure transmitting
media. The gasket use increased the limit of achievable pressures
as the result of a lower pressure gradient. The second important
step in the transformation of diamond anvils into a working tool
was an invention of a method for measuring pressure using ruby
luminescence.[12]

Different versions of diamond anvil cells are shown in Fig. 23.
Early designs of diamond anvils included as necessary elements a
device for adjusting anvils and lever mechanisms for loading them,
as shown in the figure. At present, it is preferred to provide plane-
parallelism of the working surfaces of anvils with precision machining,
and loading should be carried out using a purely screw mechanism
or a membrane device without using levers (see Fig. 23). The latter
makes diamond anvils very compact devices and allows them to be
combined with a variety of measuring equipment.

Figure 24 shows two types of diamond anvils for optical studies, developed by A. P. Novikov at the Institute of High-Pressure
Physics, RAS. Both cells are adapted for filling with helium, used as
a pressure-transmitting medium. The procedure of helium filling can
be described as follows. A gasket containing the sample and the pressure sensor (ruby) is installed in the cell. The gasket is slightly loaded
with the help of the locking nut (4), and the cell is installed in the
appropriate charger (see **Gas Charging of Diamond Anvils**), the
pressure in which is created using an external source. After creating
a pressure of about 0.2 GPa, the diamonds (1) are loaded by rotating
the locking nut (4), the gasket is deformed plastically and captures
compressed helium. Next, the cell is removed from the charger, the
pressure inside the gasket is fixed by the nut (4), and the cell is
released from the gas charging mechanism. Then the cell is provided

---

[12]Forman R. A., Piermarini G. J., Barnett J. D. and Block, S., *Science*, 176, 284
(1972).

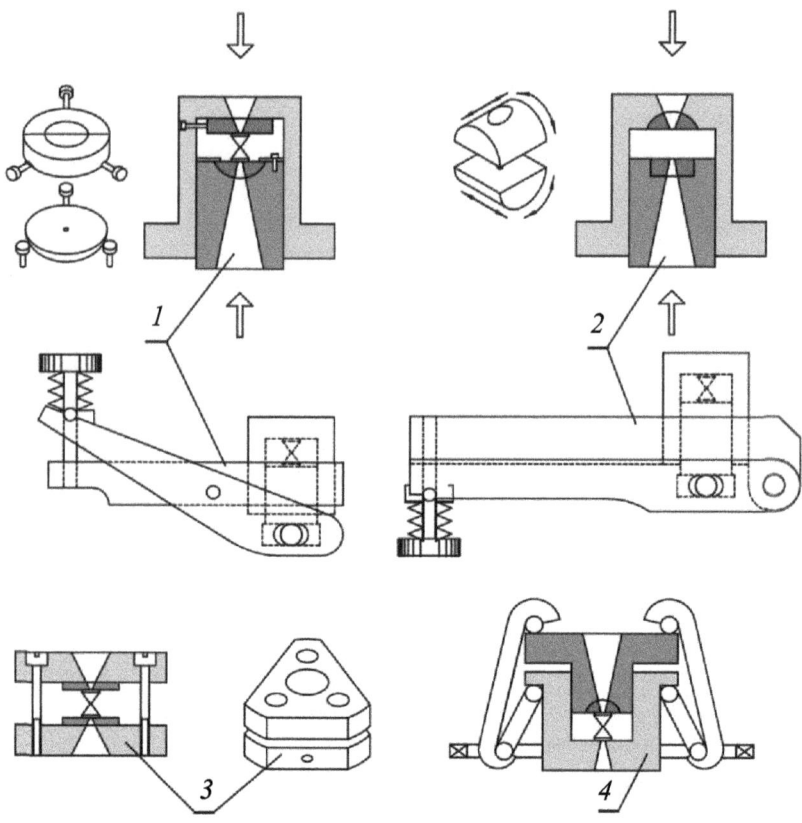

Fig. 23.   Designs of diamond anvils: 1 — cell proposed by Piermarini and Block (National Institute of Standards, USA), 2 — cell of Mao and Bell (Geophysical laboratory, USA), 3 — cell of Merrill and Basset (Rochester, USA), 4 — Holsapfel cell (Stuttgart, Germany).

with devices that drive the piston (3), whether it is a screw mechanism (5, 7, 9, 10) Fig. 24(a), or a membrane (5, 6, 7) Fig. 24(b).

Recently, a toroidal design of diamonds for anvils has been suggested in[13] that probably increased the static pressure limit, which could be achieved experimentally.

---

[13]Dewaele A., Loubeyre P., Occelli *et al.*, *Nat. Com.*, 9, 2913 (2018).

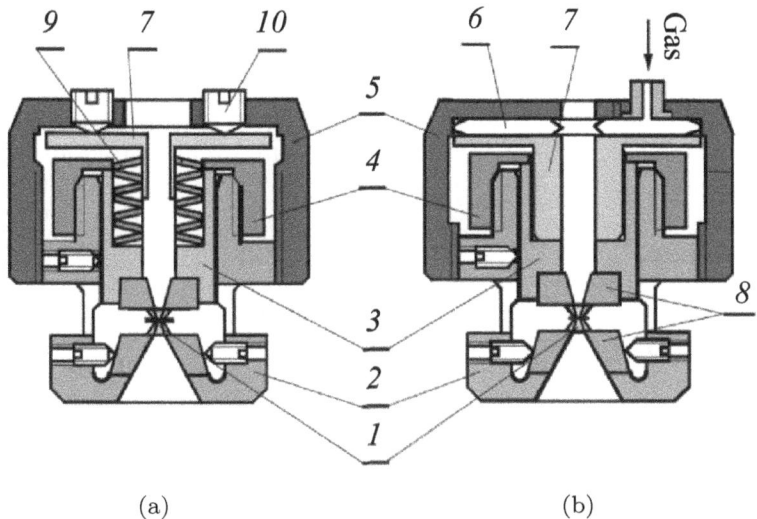

(a)   (b)

Fig. 24.   A schematic view of two types of compact diamond anvils developed by A. P. Novikov at the Institute of High-Pressure Physics of the Russian Academy of Sciences: (a) mechanically driven cell, (b) membrane cell fed by compressed helium: 1 — diamonds, 2 — body, 3 — piston, 4 — locking nut, 5 — nut, 6 — double membrane, 7 — pusher, 8 — thrust bearings, 9 — spring, 10 — power screws.

Diamond High-Pressure Scale

The diamond high-pressure scale, proposed in Ref.[14], was devoted to the study of the equation of state and Raman scattering of diamond. The analysis of the dependence of the diamond density and Raman scattering on pressure led the authors to the conclusion that the results obtained could not be consistent with the data of ultrasonic measurements and theoretical calculations without a significant change of the high-pressure scale. A new high-pressure scale was suggested, based on the diamond state equation (see Ref.[14]).

---

[14] Alexandrov I. V. *et al.*, *JETP*, 93, 680 (1987).

Diamond Superconducting

At the reaction of boron carbide B_4C and graphite at high pressures ~ 100 kbar and temperatures of 2500–2800 K, a diamond is formed with a boron content of about 3 %, which reveals metallic properties and superconductivity at $T_c \approx 4$ K.[15] The superconductivity of boron-doped diamond was confirmed on samples grown by the CVD technique.

Dilatometer

A dilatometer is a device for measurement of the length of the sample and its evolution is a function of external parameters. However, if the external parameter is pressure, then such a device is called a piezometer (see **Piezometer**). There are various designs of dilatometers. Figure 25 schematically shows one of them: a dilatometer–piezometer

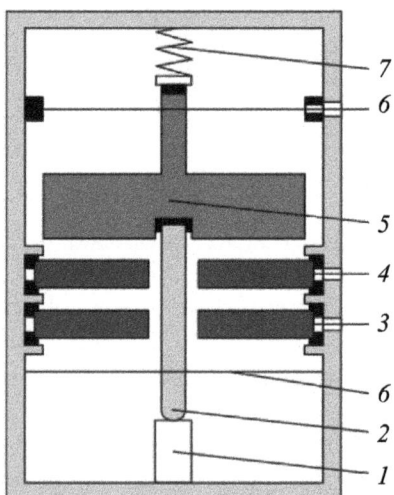

Fig. 25. Scheme of capacitance dilatometer: 1 — sample, 2 — pusher, 3, 4 — stationary electrodes, 5 — moving electrode, 6 — centering membranes, 7 — spring.

---

[15]Ekimov E. A., Sidorov V. A. *et al.*, *Nature*, 428, 542 (2004).

with a capacitance displacement sensor used to study phase transitions at high pressures in a medium of compressed helium. As can be seen from Fig. 25, the capacitance of the capacitor with electrodes 4 and 5 controls a deformation of the sample, while the capacitance of the fixed capacitor with electrodes 3 and 4 is the reference and allows to take into account the change in the dielectric constant of the medium.

Drickamer

Harry George Drickamer was an American scientist who worked in the field of high-pressure physics and chemistry (Fig. 26). He was the first to receive the P. W. Bridgman gold medal, established by the International Association for the Advancement of High-Pressure Science and Technology (AIRAPT) in 1977.

Fig. 26. Harry Drickamer (1918–2002) (from archive of the University of Illinois).

   Drickamer made a significant contribution to experimental techniques at high pressures including IR spectroscopy, Mossbauer spectroscopy, luminescence, X-ray diffraction and electrical measurements.[16] The device capable of generating pressures of hundreds of kilobars developed by Drickamer, even before the diamond anvil era, deserves special attention (Fig. 27).

Dynamic Pressure

Dynamic pressures are pressures arising from shocks, explosions (including nuclear), effects of high-power laser radiation, etc. Depending on the technique used and the test substance, the pressure

---

[16]Drickamer H. G. and Balchan A. S., High pressure optical and electrical measurements, in *Modern Very High Pressure Techniques*, edited R. H. Wentorf, Jr., Washington: Butterworths (1962).

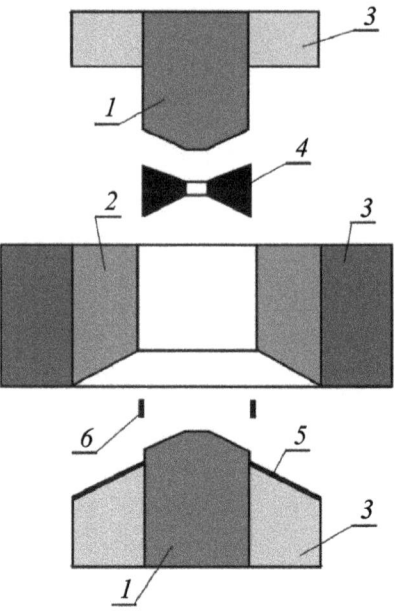

Fig. 27.   Drickamer cell for measuring electrical resistance: 1, 2 — pistons and a cell made of hard alloy, 3 — steel supporting rings, 4 — pyrophyllite, 5 — bakelite insulation, 6 — mica.

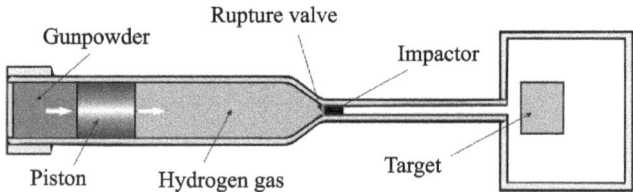

Fig. 28.   Light-gas gun.

in the shock wave can reach many hundreds of GPa and tens of thousands of degrees. Such high temperatures create certain difficulties in the analysis and interpretation of experimental data and make it impossible to investigate a substance in a condensed state.

In order to avoid excessive temperature rise, quasi-entropic compression methods are applied, which is achieved by using multiple reverberations of the shock wave in the case of explosive techniques,

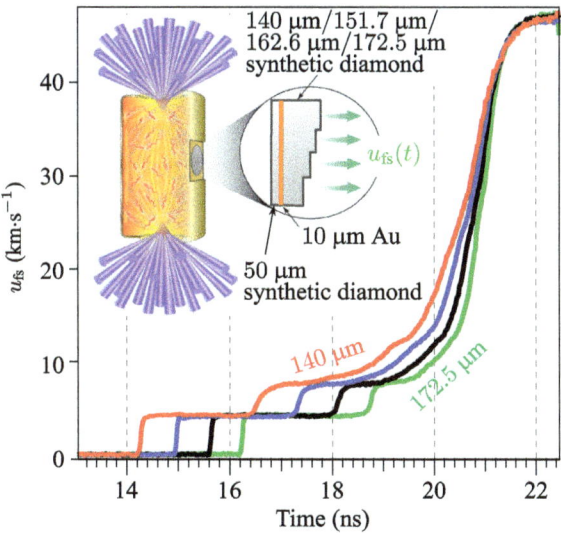

Fig. 29. The dependence of the velocity of the free surface (Ufs) of the target on time as a result of profiled action of 176 laser beams for 20 ns, which creates "non-shock" (ramp) loading up to 5TPa (see details in Smith R. F. *et al.*, *Nature*, 511, 330 (2014)).

and in the light-gas gun method (Fig. 28). The profiled laser pulse (see **Lasers and Dynamic Pressures**) (Fig. 29) or current pulse in the Z-pinch experiments produce a relatively slow ramp compression and keep the temperature within reasonable limits and get higher pressures.

$$\boxed{\text{E}}$$

Earth's Core

The Earth's core lies at a depth of 2,990 km below the surface. There is a liquid outer core, which does not transmit transverse seismic waves, and a solid inner core with a radius of 1,300 km. The pressure at the boundary of the core is ~140 GPa, in the center of the Earth the pressure is ~375 GPa. Using an analogy with the chemical compositions of meteorites, it is believed that the core of the Earth consists mainly of iron and nickel with an admixture of light elements.

Earth's Mantle

The Earth's mantle is located at depths of 30–2,900 km between the Earth's crust and the Earth's core (see Fig. 30). In the mantle, the upper mantle A (30–400 km), transition zone C (400–600 km) and lower mantle B (600–2,900 km) are distinguished. The transition zone C is characterized by an abnormally rapid increase in the speed of seismic waves and density (Fig. 30). It is believed that phase transitions occur in the transition zone, transforming terrestrial rocks and minerals to dense phases of great depth (see **Birch, Olivine–Spinel Transition, Coesite, Stishovite, Perovskite**).

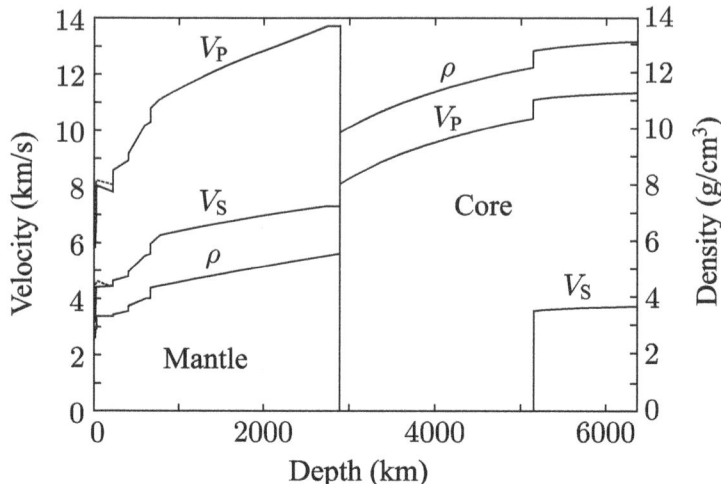

Fig. 30.   Seismic velocity and density in the Earth.

Elasticity

Elasticity is the property of a substance to return to its original state
after deformation. Note that a substance subjected to hydrostatic
compression has in some sense infinite elasticity, it always returns to
its original state. However, in the case of uniform stretching, even if
this can be done, it is not. The substance will inevitably lose continuity and break. This situation reflects the fundamental properties
of interparticle interactions, which determines the impenetrability of
particles (infinite repulsion at close distances) and the finite radius
of attractive forces. Note that the latter does not hold in the case of
quarks! For the elastic properties of crystals.[1]

Electrical Leads

Electrical leads are necessary elements of many high-pressure apparatuses. In Fig. 31, I demonstrate several ways to bring electrodes
into hydrostatic cells. In my practice, I used multichannel electrical

---

[1]Nye J. F., *Physical Properties of Crystals*, Oxford, Oxford University Press
(1985).

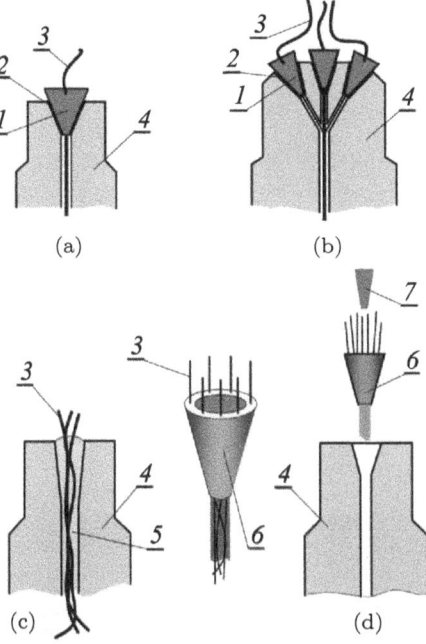

Fig. 31. Electrical lead-throughs for hydrostatic cells: (a) single-channel lead-through using the reversed Amaga cone (see **Amaga**), (b) multichannel lead-through of Amaga, (c) multichannel lead-through on epoxy resin, (d) multichannel lead-through molded from epoxy resin: 1 — metal cone, 2 — insulation, 3 — electrical wires, 4 — body, 5 — epoxy resin, 6 — molded epoxy-resin hollow cone with wires, 7 — metal cone for fitting the hollow cone (6) into the electrical lead housing.

lead-throughs (b) with the Kraft thick wrapping paper insulation up to pressures of ∼30 kbar in a liquid medium. Insulation cones made of pyrophyllite work reliably in liquid and gas media as well. The electrical lead-throughs (c) with the epoxy resin insulation are widely used for low-temperature applications in the clamp hydrostatic cells (see **High-Pressure Enclosure**), where the medium transmitting pressure is liquid. The use of these electrical lead-throughs with compressed helium at low temperatures has some specific requirements.[2]

---

[2]Petrova A. E. and Stishov S. M., *Instruments and Experimental Techniques*, 49, 592 (2006).

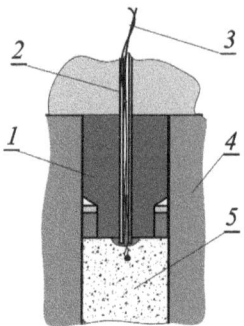

Fig. 32.   Inserting a thermocouple using a ceramic tube into a cylinder–piston cell with a solid medium transmitting pressure: 1 — plug, 2 — ceramic multichannel tube, 3 — thermocouple, 4 — cell body, 5 — medium transmitting pressure (pyrophyllite).

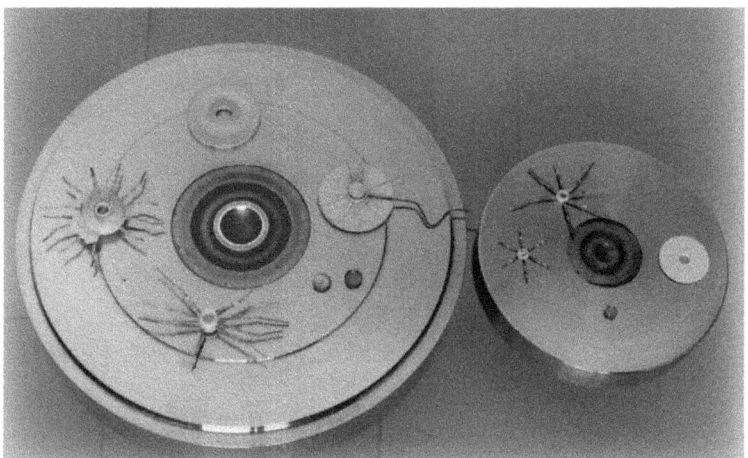

Fig. 33.   Electrical inputs to the toroid cell.

The electrical lead-throughs into devices with a solid medium transmitting pressure are performed using ceramic multichannel tubes, as shown for the case of a cylinder–piston system in Fig. 32. In the case of anvil devices, electrical leads (thermocouples) are inserted through a gap between the anvils (Figs. 7 and 33). One of the methods for placing electrodes into diamond anvils is shown in Fig. 34.

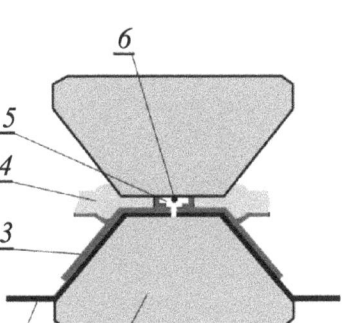

Fig. 34. Electrical inputs in diamond anvils: 1 — diamonds, 2 — deposited metal electrode, 3 — alumina (insulation), 4 — metal gasket, 5 — sample, 6 — pressure sensor.

Elements of High-Pressure Installations
(See **Bourdon Manometer, High-Pressure Enclosure, High-Pressure Tubing, Intensifier, Manganin, Piston, Press Hydraulic, Pump, Receiver, Seal, Separator, Valve, Windows Optical, etc.**)

An extremely important part of working at high pressures is to ensure the leak-free fittings of all parts of the installations. This is of particular importance in experiments with gases, and especially with helium.

My experience shows that, for example, by ensuring the tightness of the seals when working with a liquid, you find that the same seals leak when you try to compress argon, having achieved tightness against argon, you will find that your system does not "hold" helium. Figure 35 shows examples of some fittings satisfactorily operating at pressures of 10 kbar and above in a medium of compressed helium. Figure 36 illustrates a method of connecting a high-pressure tubing of 1.6 mm in diameter with the installation. The main element of the connection is the sleeve (2), which firmly squeezes the tubing (1) up to plastic deformation under loading through the sleeve (3) and the nut (4). The sleeve (3) is split, which allows to remove the tubing from the connector through the hole in the nut (4). Before assembly, the tubing (1) and the sleeve (2) must be degreased. It is useful to rub the junction of the tubing (1) and sleeve (2) with iron oxide, which increases its adhesion (see **Iron Oxide Fe$_2$O$_3$**).

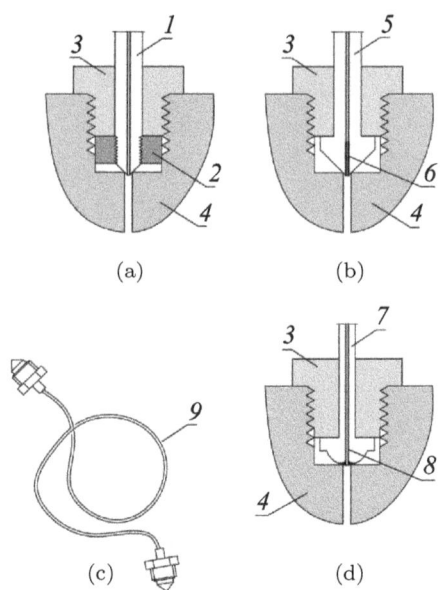

Fig. 35. Examples of high-pressure fittings: 1 — tubing, 2 — sleeve, 3 — nut, 4 — body, 5, 7 — choke, 9 — flexible tubing; (a) tubing (1) is directly pressed to the saddle by means of a sleeve (2) and a nut (3), (b) tubing is connected to the choke (5) in a hot condition with a brazed thread (6), (c) connecting parts of the installation with a flexible tubing, (d) tubing (8) is flared in place pressing the fitting (7) to the saddle.

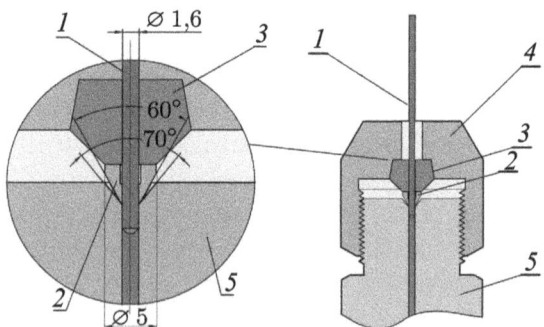

Fig. 36. Connection of high-pressure tubing with installation: 1 — tubing, 2 — sealing sleeve, 3 — pressing sleeve, 4 — nut, 5 — mounting part.

Equation of State

An equation of state is a relation connecting pressure P, volume V and temperature T. The search for a universal equation of state equally suitable for describing all substances in any state of aggregation has been undertaken repeatedly and each time has led to a disappointing result. We can describe a substance in an extremely dilute state, that is, in an ideal gas state, or a substance in an extremely compressed state, for example, the white dwarf matter (see **White Dwarfs**), where only Coulomb attraction and Fermi repulsion are important. In the latter case, everything can be calculated, knowing only the fundamental physical constants. However, everything that interests researchers is in the interval between these two cases.

In this area, the semi-empirical Birch–Murnaghan state equation (see **Birch**) is successfully used to approximate and interpolate experimental and calculated data in a broad pressure range. Birch also used this equation to extrapolate experimental data to the pressure range corresponding to the Earth's mantle (see **Earth's Mantle**). The Birch–Murnaghan equation of state is obtained by expanding the free energy into a Taylor series in degrees of deformation (small parameter) $f = [(V_0/V)\,2/3 - 1]\,/2$ and has the form:

$$P = 3K_0/2\big[(V_0/V)^{7/3} - (V_0/V)^{5/3}\big]\{1 + 3/4(K_0' - 4)[(V_0/V)^{2/3} - 1]\},$$

where K_0 is the bulk modulus, V_0 is the volume, $K_0' = dK/dP$ at $P = 0$.

Obviously, at very high pressures, the strains cease to be small, which makes the corresponding expansion not valid. This and similar equations do not contain temperature. The latter can be taken into account using the Mie–Gruneisen equation:

$$P = P_0 + \gamma E_t/V,$$

where P_0 is the pressure at $T = 0$, E_t is thermal energy, V is the volume, and γ is the Gruneisen constant. In the classical limit, the equation takes the form:

$$P = P_0 + \gamma 3RT/V.$$

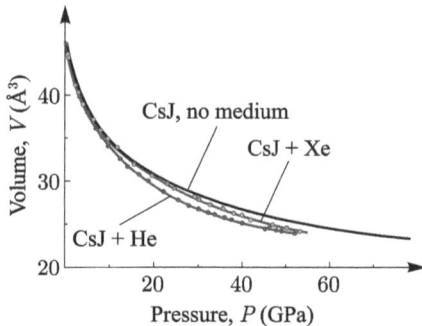

Fig. 37. Dependence of CsJ volume on pressure in various pressure transmitting media.

Useful information on the equation of state topic can be found in publications.[3] We also note that the results of the experimental study of the "equation of state" of a substance depend on the properties of the pressure transmitting medium (Fig. 37), which should be taken into account when interpreting the experimental data.[4]

European High-Pressure Research Group (EHPRG)

The EHPRG was established in 1963, mainly through the efforts of S. Goodman, while an employee of Standard Telecommunications Laboratories, Harlow (United Kingdom). The EHPRG organizes annual European high-pressure conferences. In 1989, a prize for young scientists was established at the conference in Paderborn (Germany).

[3]Stacey F. D., Brennan B. J. and Irvine R. D., *Geophysical Surveys*, 4, 189 (1981).

[4]On this subject, see Avilov V. V. and Arkhipov R. G., *Sol. State. Comm.*, 48, 657 (1963).

$$\boxed{\textbf{F}}$$

Finite Elements Method

The finite elements method is a technique of determining numerical solutions of partial differential equations, which allows solving many problems in electrodynamics, hydrodynamics, mechanics, etc.[1]. There are a number of software packages suitable for calculating the strength of high-pressure apparatuses and devices.

Fluorinert

Fluorinert is a fluorocarbon liquid used in cooling electronic devices. It is also used as a pressure transmitting medium in neutron studies at high pressures, since it does not contain hydrogen, a strong neutron absorber.

Friction

Friction prevents movement of one body relative to another. In the technique of high pressures, we are faced with cases where it is desirable to make as little friction as possible or as much as possible. The first case is friction in the piston–cylinder devices. Here, friction absorbs a part of the force and can lead to cold welding in

---

[1]Zienkiewicz O., Taylor R. and Zhu J. Z., *The Finite Element Method: Its Basis and Fundamentals* (7th edn.), Butterworth-Heinemann (2013).

some places of the piston pair. To avoid this, the surfaces of the pair should be carefully processed. It is desirable that the elements of the pair are made of different materials: from different steels, steel and bronze, hard alloy and steel, etc. It is best to make anti-extrusion rings used in seals from beryllium bronze (having anti-friction properties) and coated with indium. The second case is the friction of the pressure transmitting medium in multi-anvil installations, toroids and lentils. Here, it is necessary to prevent excessive flow of material into the gap between the anvils, otherwise, you will not be able to get high pressure. The materials used for these devices: **Pyrophyllite**, **Lithographic Stone** have the necessary frictional properties when sliding on a hard alloy. Sometimes corresponding surfaces are coated with iron oxide to increase friction (see **Iron Oxide Fe$_2$O$_3$**).

$$\boxed{\text{G}}$$

Gas Charging Diamond Anvils

Filling diamond anvils with a substance that is in a gas state at normal pressure and room temperature (for example, inert gases, hydrogen, etc.) requires a certain skill. Historically, a cryogenic method was used for this purpose, when the entire diamond cell with a prepared gasket containing the sample and pressure sensor is immersed in a cryostat with a liquefied gas. At the same time, there should remain a gap between the anvils and the gasket sufficient for the penetration of fluid into the cell. Next, the cell is closed, using a corresponding "crank", and removed from the cryostat. Currently, a slightly different method of filling cells is used more often, as shown in Fig. 38. Here, a diamond anvil cell devoid of a power drive is placed in a high-pressure vessel. Compressed gas is supplied to the vessel using a membrane compressor or other device that ensures the purity of the gas. The remaining procedures are similar to those described for the case of cryogenic filling.

Gasket

A gasket is a metal sheet placed between diamonds in the technique of diamond anvils. The gasket has a hole, which can be filled with a pressure medium, a sample and a pressure gauge such as a ruby chip.

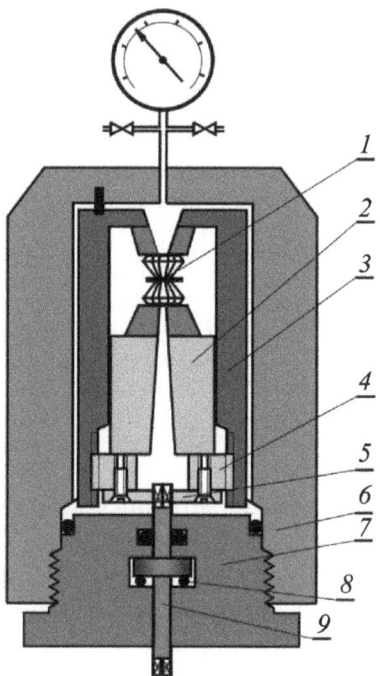

Fig. 38. Device for filling diamond anvil cells with gaseous medium: 1 — diamond anvils, 2, 3 — piston and cylinder of diamond cell, 4 — fixing nut, 5 — disk with crank nest, 6 — high-pressure vessel, 7 — plug, 8 — bearing, 9 — crank.

Gigapascal

The gigapascal is the SI unit of pressure:

$$1\,\mathrm{GPa} = 10\,\mathrm{kbar} = 10^{10}\,\mathrm{dyn/cm^2}.$$

Girdle

The girdle high-pressure apparatus has a design similar to the "Belt" apparatus.[1] An analogous construction is described in Ref.[2].

---

[1]Wilson W. B., *Rev. Sci. Instr.*, 31, 331 (1960).
[2]Daniels W. B. and Jones M. T., *Rev. Sci. Instr.*, 32, 885 (1961).

Fig. 39. Meriden, New Hampshire, the school building where the first high pressure Gordon conference was held.

Gordon Conferences on High-Pressure

The first conference, which marked the beginning of a series of biannual conferences, called Gordon Conferences, was organized by Professor Neil E. Gordon at John Hopkins University in 1931. The purpose of these conferences was and is to discuss the most important issues of science among professionals working at its advanced frontiers.

From 1931 to 1947, these conferences were held in various places in Maryland. In 1947, the location of the conferences moved to New Hampshire, and the conferences themselves were named Gordon's in honor of their founder. Currently, Gordon conferences are held in many US states and even in other countries.

The first Gordon High Pressure Conference was held in 1955 at Meriden, New Hampshire, in a school building that was free in the summer (see Fig. 39). Chairmen and organizers of the Gordon High Pressure Conference at various times were H. Drickamer, O. Anderson, S. Swenson, W. Paul, J. Jamieson, N. Ashcroft and other high-pressure pioneers. At the conference, young scientists are awarded the Jamieson Prize and the Van Valkenburg Prize.

Graphene

Graphene is a two-dimensional crystal consisting of one atomic layer of carbon. It was first obtained by splitting graphite crystal.

Graphene has unique mechanical and electronic properties and is widely used in electronics and other applications.

Graphite

Graphite is a layered black carbon phase that conducts electricity. The crystal structure consists of layers of carbon atoms that form the structure of the honeycomb type (Fig. 40). Separate layers are interconnected by weak van der Waals bonds, which determine the lubricating properties of graphite and its use in the production of various pencils. Graphite has extensive industrial applications as a variety of electrodes, current collectors, heat-resistant materials, neutron

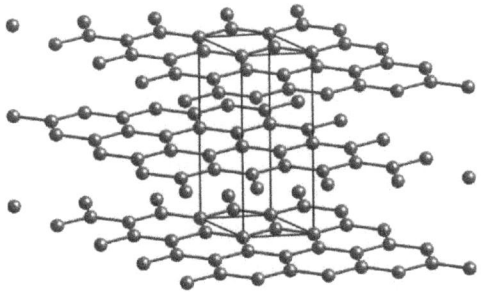

Fig. 40.   Graphite crystal structure.

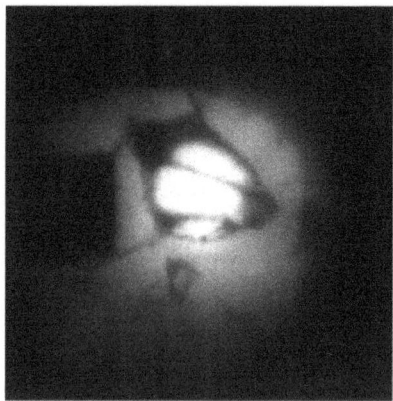

Fig. 41.   Transparent graphite at high pressure.

moderators. It is widely used in the technique of high pressure for the manufacture of short-circuited heaters.

At high pressures and temperatures ($\sim$55 kbar, $1,500°$C), graphite turns into diamond (see **Diamond**). However, when graphite is compressed at room temperature, it goes into an amorphous semi-transparent state (Fig. 41).

H

Hall

Howard Tracy Hall was an American physical chemist educated at the University of Utah at Salt Lake City (Fig. 42). In 1948, he received the PhD degree, under the guidance of Prof. Henry Eyring. In the same year, he joined the General Electric Research Laboratory, a group whose main task was to develop a method for growing artificial diamonds. The thing was that the manufacture of tungsten spirals for the production of electric lamps, produced in the General Electric Company required the use of expensive diamond dies, which made the company management set the appropriate task. It was in this laboratory that Hall with minimal support built the "Belt" apparatus[1] (see **Belt**), with which the first artificial diamonds were obtained.[2] Shortly after the successful synthesis of the diamond, Tracy Hall left General Electric and was promoted to Professor at the Brigham Young University. However, the government forbade him to use his "Belt" for reasons of secrecy.

Fig. 42. Tracy Hall (1919–2008).

[1] Hall H. T., *Rev. Sci. Instr.*, 31, 125 (1960).
[2] Bovenkerk H. P., Bundy F. P., Hall H. T., Strong H. M. and Wentorf R.H., Jr., *Nature*, 184, 1094 (1959).

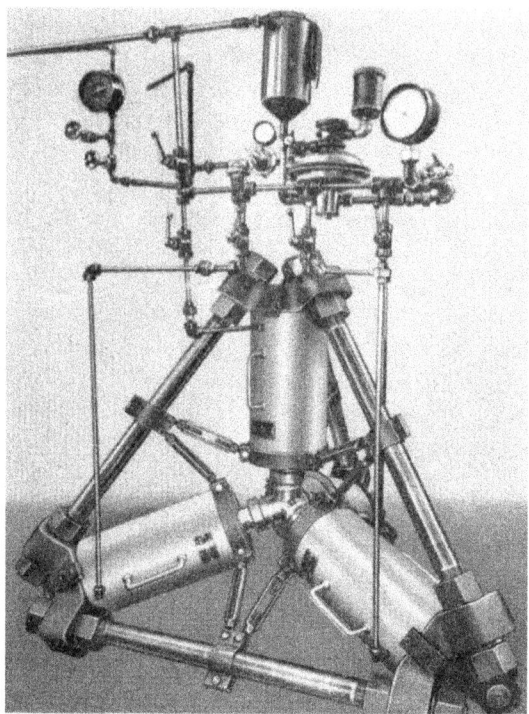

Fig. 43.   The first tetrahedral setup created by Hall H. T.[3]

He had no choice but to invent a new high-pressure apparatus, which he successfully did, having successfully constructed a working tetra-hedral press[3] (Fig. 43).

Hard Alloys

Hard alloys are a group of hard materials consisting mainly of carbides of tungsten, some times with addition of carbides of titanium,

---

[3]Hall H. T., *Rev. Sci. Instr.*, 29, 267 (1958).

tantalum and others. These materials contain cobalt as a binder. The compressive strength of individual grades of these cemented carbides may exceed 600 kg/mm^2. That is why hard alloys are used for fabrication of various kinds of pistons and other components of high-pressure equipment. The hard alloy (Carboloy) consisting of tungsten carbide, cemented with cobalt (WC+Co), was widely used by P. Bridgman in his high-pressure experiments (see **Bridgman Anvils, Lentils, Toroid**).

Heat Treatment

Heat treatment is a procedure associated with thermal effects on the material in a certain mode (quenching, aging, annealing, etc.) to obtain the desired properties.

Helium

Helium, the second element of the periodic table of Mendeleev, was discovered in 1868 during a spectral study of the Solar corona. In 1881, helium was found on Earth in volcanic gases. Liquid helium was obtained in 1908 by the Dutch physicist H. Kamerling-Onnes, which subsequently allowed him to discover the phenomenon of superconductivity. Helium has two stable isotopes, ^{3}He and ^{4}He. The boiling point of ^{4}He at normal pressure is only 4.215 K. Solid helium was obtained by V. H. Keesom in 1926 under pressure at $\sim$35 atm. Keesom also discovered in 1932 the so-called λ-transition in liquid helium, which, as shown by P. L. Kapitsa in 1938 (and independently by Allen and Mizener), turned out to be a transition of helium into a superfluid state. The theory of superfluidity was built by L. D. Landau in 1941 and R. Feynman in 1945.[4]

Liquid helium is the most important refrigerant, without which the very existence of low-temperature physics would be impossible. The creation of superconducting magnets of modern accelerators,

---

[4]Mendelssohn K., *The Quest for Absolute Zero*. New York: McGraw-Hill (1966).

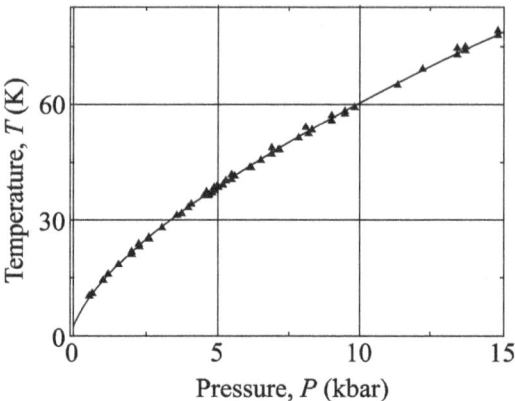

Fig. 44.   The dependence of the melting point of helium ^{4}He on pressure.

thermonuclear installations, medical diagnostic devices, etc. would be impossible without liquid helium. Helium also plays a large role as a medium providing hydrostatic conditions in experiments at high pressures carried out with diamond anvils (see **Diamond Anvils**) and at low temperatures. Figure 44 shows that at a boiling point of liquid nitrogen of 77 K, the liquid helium medium provides ideal hydrostatic conditions up to 15 kbar. Note that at room temperature, helium solidifies at a pressure of ~120 kbar, which creates the hydrostatic and quasi-hydrostatic conditions when helium is used as a medium transmitting pressure in diamond anvils. Currently, the melting curve of helium is known up to temperatures of 600 K.

High-Pressure Enclosure

A high-pressure enclosure, as a rule, is a strong vessel capable of withstanding a certain pressure, which may contain electrical lead-throughs and optical windows, or be made of materials that are "transparent" for neutrons and X-rays, or non-magnetic materials. High-pressure enclosures can be autonomous or connected to a pressure generator. The currently popular diamond anvils (see **Diamond Anvils**), apparently, can not be classified as "vessels", although they can be called cells. The same applies to the various anvil systems (see **Bridgman Anvils, Lentils, Toroid**), hydrostatic cells,

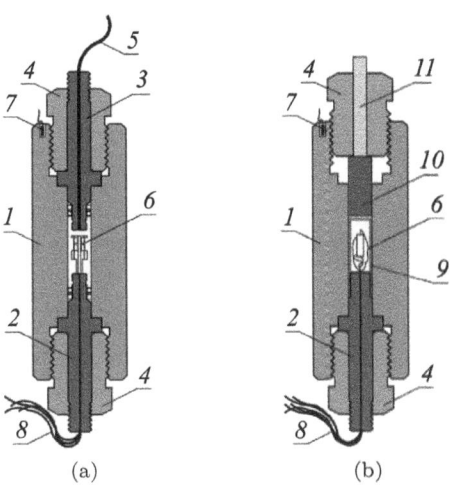

Fig. 45. High-pressure cells to study the physical properties of a substance at low temperatures: (a) cell for studies in a medium of compressed helium, (b) cell of fixed pressure or clamp cell for studies in a medium of frozen liquid: 1 — body, 2 — electrical input, 3 — input of compressed helium, 4 — nuts, 5 — tubing, supplying compressed helium, 6 — samples, 7 — temperature sensor, 8 — wires, 9 — teflon sealing cup, 10 — piston, 11 — pusher.

connected with the source of pressure (see **High-Pressure Installation (Hydrostatic)**), multi-anvil installations (see **Multi-anvil Installations**), etc. Figure 45 schematically shows two variants of enclosures or cells for low-temperature studies. The cell (a) is connected by a flexible tubing (5) to a compressed helium generator. In the case of (b), pressure in the enclosure, filled with liquid, is created by an auxiliary press. The latter moves the piston (10) into the cell by means of the pusher (11). Upon reaching the required pressure, the position of the piston is fixed with the help of a nut (4). The compressed fluid is sealed with a Teflon cup (9). The cell (b) is usually called the "clamp cell".

High-Pressure Installation (Hydrostatic)

Figure 46 shows a schematic view of a high-pressure installation with variable mechanical support based on a hydraulic press. When using a hard alloy piston, pressures of $\sim$30 kbar in the medium of pentane or gasoline are easily achieved in the installation. Figure 47 shows a

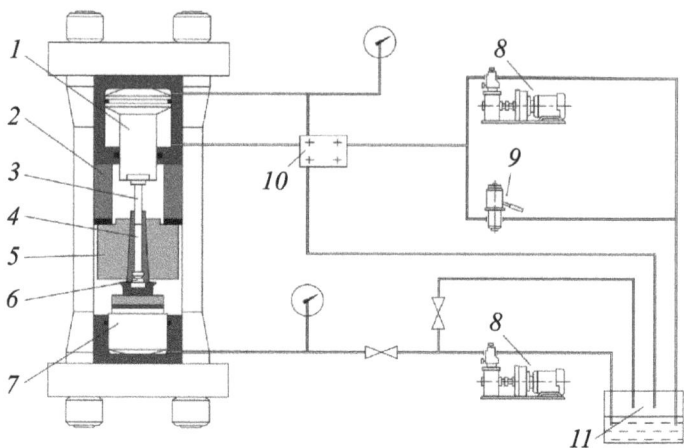

Fig. 46. Schematic view of a hydrostatic installation with mechanical variable support based on a two-cylinder hydraulic press (in fact, this installation is an almost complete reproduction of the 30,000 Bridgman installation): 1 — main cylinder piston, 2 — thrust ring, 3 — piston of high-pressure enclosure, 4 — high-pressure enclosure, 5 — supporting ring, 6 — plug with electric inlets, 7 — piston of mechanical support cylinder, 8 — mechanical pump, 9 — manual pump, 10 — valves, 11 — hydraulic liquid tank.

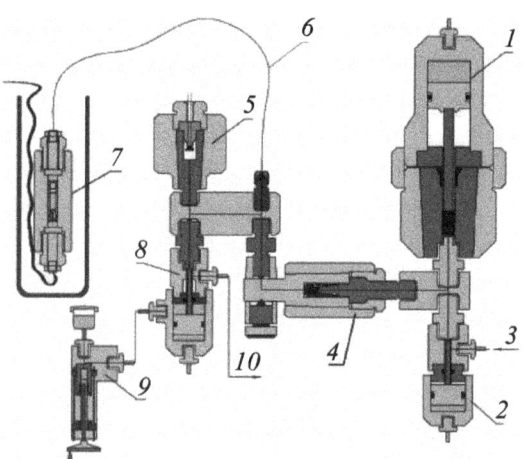

Fig. 47. Schematic view of installation for creating hydrostatic pressures in a gas or liquid medium: 1 — intensifier, 2 — hydraulically controlled valve, 3 — input for compressible medium (liquid or gas), 4 — high-pressure valve, 5 — manganin gauge bomb, 6 — high-pressure tubing, 7 — vessel, containing the sample and necessary inputs, 8 — relieve valve, 9 — screw press controlling valve 8, 10 — outlet for a relieved medium.

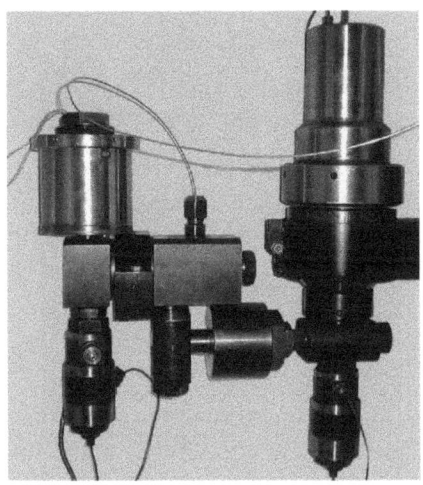

Fig. 48.   View of the main components of the installation shown in Fig. 47.

scheme of a universal high-pressure unit, which allows compression of gases and liquids. The main feature of the installation is an absence of rigid connection between the vessel and the pressure generator. As shown in Figs. 47 and 48, the vessel is connected to the intensifier by means of flexible tubing and can be placed in a cryostat or a furnace. The pressure limit in such an installation is determined by the strength of the high-pressure tubing and is about 15–18 kbar.

High-Pressure Installation (Solid Medium Transmitting Pressure)

A high-pressure installation with a solid medium transmitting pressure is not much different from the installation shown in Fig. 46. For example, the high-pressure enclosures of the type of **Belt, Lentil** or **Toroid** can be used in conjunction with a single-cylinder hydraulic press (see **Press Hydraulic**) and with the hydraulic system shown in Fig. 46.

The only difference of this type of installation is a presence of low-voltage power source for short-circuited heaters and high-power copper buses supplying high-ampere current to the enclosure. A block diagram of such an installation is shown in Fig. 49.

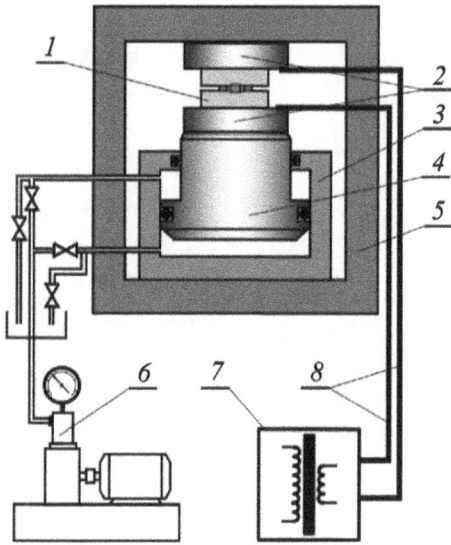

Fig. 49. Block diagram of an installation with a solid medium transmitting pressure: 1 — high-pressure cell, 2 — thrust block, 3, 4 — working cylinder, 5 — press frame, 6 — pump, 7 — low-voltage power supply, 8 — copper buses.

High-Pressure Tubing

High-pressure tubing are thick-walled metal tubes used in high-pressure equipment for connecting the parts. They are made of steel, beryllium bronze, brass and copper. Depending on the ratio of external and internal diameters, material and heat treatment, high-pressure tubings can work up to ~ 18 kbar. In my practice, I used tubings made of stainless steel with a diameter of 3.1 (0.5) mm and 1.6 (0.2) mm with a tensile strength of 16 and 12 kbar, respectively (the internal diameter is shown in parentheses).

Hugoniot

The Hugoniot adiabat is a relationship between the volume and pressure behind the shock wave front, which follows from the equation relating the thermodynamic quantities before and after passing the shock wave (Hugoniot–Rankin equation). The entropy of a substance increases along Hugoniot adiabat.[5]

Hydrides

Hydrides are hydrogen compounds that acquired a new meaning after N. Ashcroft remarked about a possibility of metallization and occurrence of high-temperature superconductivity in a number of hydrides at pressures significantly lower than the pressure of hydrogen metallization (see **Hydrogen**). The idea was that the H–H distance in hydrides may be less than in pure hydrogen at equivalent pressure, which would mean a very high frequency of phonons responsible for the pairing of electrons. Ashcroft pinned his hopes on the hydrides of the elements of the fourth group, but success came in the study of hydrogen sulfide H_2S. It was shown in Ref.[6] that hydrogen sulfide subjected to a pressure of 150 GPa exhibits superconducting properties at temperatures of ~ 203 K. As it turned out later, the superconducting phase in the H–S system has the composition H_3S. An even higher temperature, apparently, of the superconducting transition was found in the study of lanthanum hydride LaH_{10} (245–280 K, 190–200 GPa).[7]

Hydrogen

Hydrogen, the number one element in the periodic table, is a gas under normal conditions. Hydrogen as a light and combustible gas has numerous applications (hydrogen burners, coolants, filling balloons and airships, rocket engines, etc.). Hydrogen was first liquefied in 1885 by Karol Olszewski (Krakow) as a result of intense rivalry with James Dewar (London). However, it was Dewar who made liquid hydrogen in appreciable quantities, thanks to the vacuum

---

[5]Landau L. D. and Lifshitz E. M., *Statistical Physics* Vol. 5 (3rd edn.). Oxford: Butterworth–Heinemann (1980).
[6]Drozdov A. P., Eremets M. I. *et al.*, *Nature*, 525, 73 (2015).
[7]Somayazulu M. *et al.*, *Phys. Rev. Lett.*, 122, 027001 (2019).

vessel he invented. The condensation and crystallization temperatures of hydrogen turned out to be equal to 20.271 K and 13.99 K, respectively.

Hydrogen is usually found in molecular form, that is, in the form of molecules consisting of two protons and two electrons. However, if you split the molecule, you get atomic hydrogen, consisting of one proton and one electron, representing a complete analogue of alkali metals. Atomic hydrogen is not stable at atmospheric pressure and quickly turns into a molecular form. However, in the 1930s, the British scientist John Bernal (see **Bernal**), a well-known fighter for peace and the author of original scientific ideas (for example, about the structure of fluids, about the olivine–spinel transition in the depths of the Earth) suggested that atomic hydrogen may be stable at high pressures. This idea attracted the attention of theorists Wigner and Huntington, who made the corresponding calculations in 1935.[8] Bernal's hypothesis was confirmed in the above calculations. According to these calculations, molecular hydrogen enters the atomic metal phase at high pressures of about 300 kbar with a significant increase in density (in the light of current research, the estimate looks very approximate). However, for a long time the problem of metallic hydrogen has not been the focus of attention of researchers in the field of high-pressure physics.

The theoretical work of A. A. Abrikosov (future Nobel laureate) and V. Trubitsin, made in the 1950s, and dedicated to the structure of the hydrogen planets: Jupiter and Saturn is worth mentioning here.

An explosion of interest in the problem occurred in 1968, when the American physicist Neil Ashcroft[9] published an article in which he argued that atomic metallic hydrogen can be stable at atmospheric pressure. But most importantly, Ashcroft showed that metallic hydrogen can have superconductivity at room temperature. This was the main motivation for further research. Indeed, obtaining a material with superconductivity at room temperature would mean a revolution in the power industry.

Several experimental groups announced their intentions to engage in the production of metallic hydrogen. Neil Ashcroft and

---

[8]Wigner E. and Huntington H. B., *J. Chem. Phys.*, 3, 764 (1935).
[9]Ashcroft N. V., *Phys. Rev. Lett.*, 21, 1748 (1968).

Arthur Ruoff of Cornell University (USA) were among the first to announce themselves. In the interview published in *Physics Today* in the early 1970s, they stated they would make metallic hydrogen within a year with sufficient funding.

However, at that time it was already clear that to produce metallic hydrogen one would need to obtain a pressure of at least a million atmospheres. At the same time, no one knew how to obtain this million. Nevertheless, there was no real progress on the way to reaching pressure in the millions until Peter Bell and Dave Mao from the Geophysical Laboratory (Washington, USA) declared in 1977 that they reached 1.7 million atmospheres in a miniature device (diamond anvils), the pressure in which is created using two diamonds. So the study of hydrogen in the megabor range of pressure started. At that time, there were four effective players in this field. These were Arthur Ruoff from Cornell, Russell Hemley and Dave Mao from the Geophysical Laboratory and Isaac Silvera from Harvard (all from the USA). Silvera was recognized for his work on the stabilization of atomic hydrogen deposited on a substrate coated with superfluid helium during his stay in Netherlands.

Over the years of research, many experimental data have been obtained under conditions of high static pressures and dynamic pressures. Bill Nellis from the Livermore Laboratory (USA) argued that liquid deuterium (a hydrogen isotope) transforms into a metallic liquid at high pressures and temperatures created by strong shock waves. Improving the technique of the experiment, the researchers, step by step, approached the pressures reigning in the center of the Earth (about 4 million bars). However, hydrogen stubbornly did not want to metallize. The researchers turned to hydrogen compounds and liquid hydrogen, many interesting results were obtained, but solid hydrogen remained uncompromising. And finally, Ike Silvera[10] published in *Science* a report on the production of metallic hydrogen at a pressure of about 5 million bars (see hydrogen phase diagram, Fig. 50).[11] Of course, this statement was instantly criticized by competing researchers, although it was supported by the father of all this activity, Neal Ashcroft. Quite recently a French team announced the

---

[10]Dias R. and Silvera I., *Science*, 355, 6326 (2017).
[11]Utyuzh A. N. and Mikheyenkov A. V., *Phys. Usp.*, 60, 886 (2017).

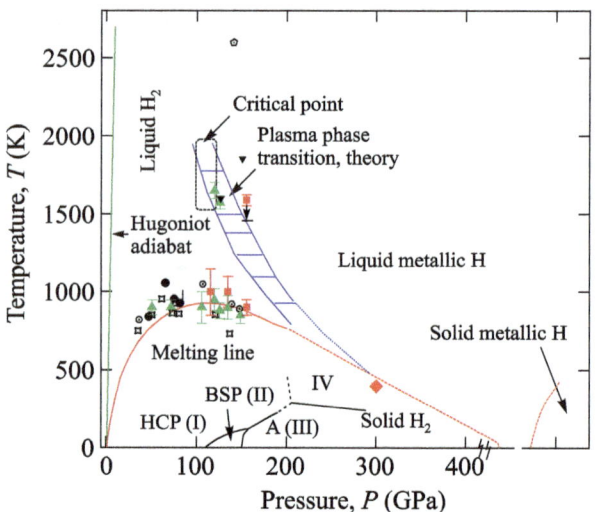

Fig. 50.   Phase diagram of hydrogen.

metallic transition in hydrogen at 425 GPa.[12] Anyway, it is clear that metallic hydrogen is obtained or will be obtained, but it is important to know that the amount of substance produced in diamond anvils can be placed on the tips of a thin sewing needle. For this reason, any projects related to the practical use of metallic hydrogen should be attributed to the science fiction genre (see the review on the problem of metallic hydrogen[13]). It should also be remembered that there are isotopes of hydrogen: deuterium and tritium, the phase diagrams of which may differ significantly in some places.

Hydrothermal Process

A hydrothermal process involves chemical reactions in an aqueous medium at temperatures above 100°C and elevated pressure. It is also widely used for laboratory and industrial growth of quartz crystals, sapphires, emeralds, etc. The hydrothermal process is implemented using high pressure autoclaves (see **Autoclave**).

---

[12]Loubeyre P., Occelli F. and Dumas P., Observation of a first-order phase transition to metal hydrogen near 425 GPa, https://arxiv.org/abs/1906.05634.
[13]Utyuzh A. N. and Mikheyenkov A. V., *Phys. Usp.*, 60, 886 (2017).

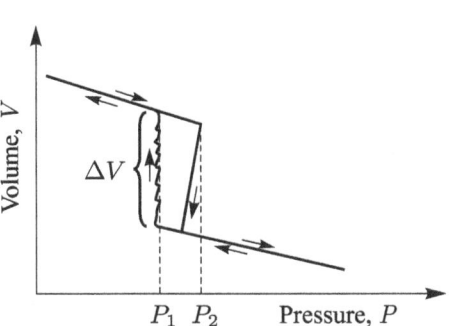

Fig. 51. Hysteresis during crystallization and melting of argon at high pressures. The downward arrow indicates spontaneous crystallization of argon under some supercooling, while the pressure drops noticeably due to a decrease in volume during crystallization. An almost equilibrium melting of argon occurs on the jagged line with the arrow pointing upwards. The experimenter each time slightly lowers the pressure in the system and waits until the pressure recovers when some fraction of the substance melts. Each clove corresponds a separate step of the described manipulation.

Hysteresis

Hysteresis is the phenomenon of delayed response of the system to the applied action. There are different types of hysteresis: magnetic, ferroelectric, elastic, etc. The phenomenon that is of interest in the framework of this book, refers to the physics of phase transitions. Indeed, during phase transitions of the first order (see **Phase Transitions**), the pressure at which a phase transition occurs during loading does not coincide with the pressure of the phase transition during unloading. In most cases, this phenomenon is determined by kinetic effects. Figure 51 shows the hysteresis that occurs when argon is crystallized at high pressure. The observed hysteresis is associated with the fact that in order to form the nucleus of the crystalline phase during the compression of a liquid, it is necessary to achieve a certain degree of supercooling or "oversqueezing". The effects of friction create a somewhat different kind of phenomenon observed at high pressures. The subsection **Piezometer** demonstrates the hysteresis that occurs when measuring the compressibility of a substance using a piston piezometer.

$$\boxed{\text{I}}$$

Ice Bomb

B. G. Lazarev and L. S. Kahn, using the anomalous properties of water to increase the specific volume during freezing, constructed a so-called ice bomb to study superconductivity at high pressures. The authors froze the water at a constant volume in a strong vessel (bomb), while the pressure reached a value of the order of 2 kbar.[1]

Indium

Indium is a soft, ductile and inert metal with a melting point of 156.6°C. It is used to seal vacuum joints. It is also utilized in the high pressure technique for the manufacture of soft gaskets. It is useful to cover sealing rings, steel or bronze with a thin layer of indium to reduce friction.

Inert Gases and Their Compounds

Inert gases are the elements of the zero group of the periodic table: He, Ne, Ar, Kr, Xe, Rn. Of these, He, Ne, Ar and sometimes Xe (see **Helium** and **Argon**) are used as pressure transmitting media. As it turned out, inert gases at high pressures form pseudo compounds of the type $He(N_2)_{11}$, $Kr(H_2)_4$, $Ar(H_2)_2$, $Ar(O_2)_3$. In fact,

---

[1] Lazarev B. G. and Kahn L. S., *ZhETF*, 14, 463 (1944).

these compounds can be considered as a geometrically ordered system consisting of solid spheres of various sizes.

Institute for High-Pressure Physics of the Russian Academy of Sciences (HPPI)

The Institute of High-Pressure Physics of the Russian Academy of Sciences located in Troitsk (New Moscow) was founded by L. F. Vereshchagin in 1958 and now bears the name of its founder. Popular high-pressure cells: lentils and toroids are created in the HPPI (see **Bridgman Anvils, Lentils, Toroid**). The research in the HPPI formed the basis for creating the industry of artificial diamonds in the USSR. Dense silica with a crystalline structure of rutile, the natural analogue of which was called stishovite (see **Stishovite**), also received a birth in the HPPI.

Intensifier

An intensifier is a device for increasing pressure using a differential piston system, as shown in Fig. 52. The pressure of the liquid or gaseous medium at the output of the intensifier 4 exceeds the pressure (multiplied) at the inlet 5 with a coefficient equal to the ratio of the areas of the pistons 1 and 2, taking into account the correction for friction. Intensifiers are an integral part of high-pressure liquid and gas installations (see **High-Pressure Installation**).

Iron

Iron, an element with atomic number 26, belongs to the group of transition metals with a non filled d-electron shell. As a structural material, iron and its alloys played a prominent role in the development of human civilization. However, the role of iron in the internal structure of the Earth and planets is also important. The fact is that the composition of meteoritic material and geophysical data indicate an existence of iron in the Earth's core (see **Earth's Core**). The Earth's core is located at a depth of 2,900 km, and the so-called external core is in a liquid state. For this reason, studies of the phase

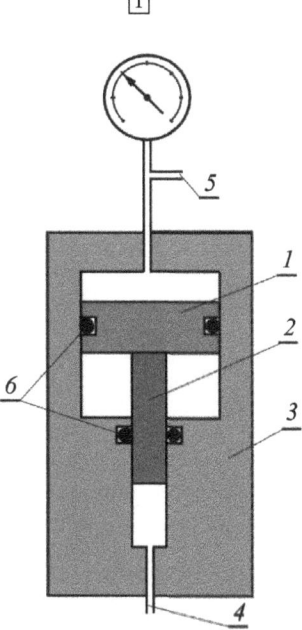

Fig. 52. High-pressure intensifier (booster): 1 — low-pressure piston, 2 — high-pressure piston, 3 — body, 4 — output liquid of increased pressure, 5 — input of working fluid (from the pump), 6 — seals.

diagram of iron are of particular importance and will ultimately verify one or other ideas about the structure and composition of the Earth's core. The pressures and temperatures prevailing in the depths of the Earth are so high (millions of atmospheres and thousands of degrees) that these studies are extremely difficult. Figure 53 shows the phase diagrams of iron constructed according to numerous static and dynamic experiments and theoretical studies.[2]

Iron Oxide Fe_2O_3

Iron oxide (Fe_2O_3) — commercially available as a brown powder — has very high frictional properties. With the right choice of parameters, a thin layer of powder can replace soldered and threaded connections, which can be used in high-pressure techniques. On the

---

[2] Anzellini S. *et al.*, *Science*, **340**, 464 (2013).

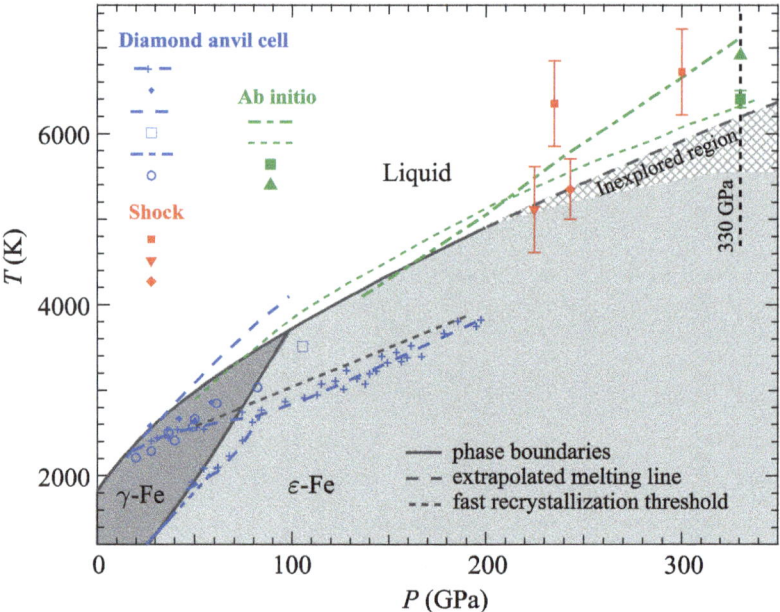

Fig. 53.   Iron phase diagram (see details in Jayaraman A., *Phys. Rev.* **137**, A179 (1965)).

other hand, iron oxide Fe_2O_3 is known in nature as the mineral "hematite".

Isostructural Phase Transitions

Isostructural phase transitions are phase transitions without changing the crystal structure of a substance. A striking example of an isostructural phase transition is the phase transition in metallic cerium Ce, discovered by P. Bridgman at pressure of about 7 kbar and room temperature. Subsequently, E. G. Poniatowski discovered a critical point on the first order phase transition line in Ce (Fig. 54) (see **Critical Point**).

Isotopic Effects

Isotopic effects in the thermodynamic properties of a substance (equations of state, phase boundaries, etc.) arise in connection with

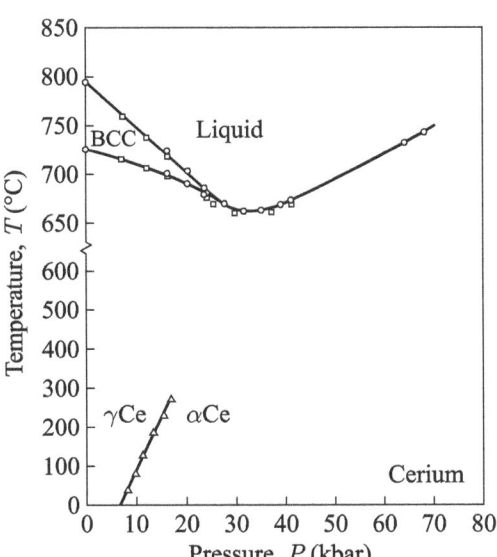

Fig. 54. Cerium phase diagram, $\gamma-\alpha$ isostructural transition terminating at a critical point is shown in Jayaraman A., *Phys. Rev.* **137**, A179 (1965).

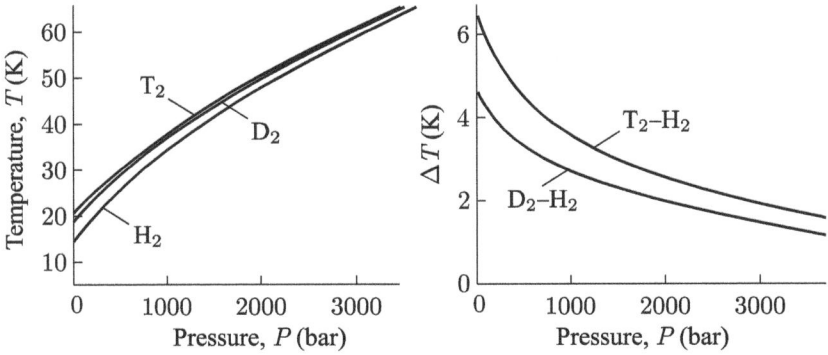

Fig. 55. Isotopic effects at melting of hydrogen and its isotopes at high pressures.

manifestations of the quantum nature of matter. Naturally, these effects are essential in systems of light isotopes, such as H, D and T, ^{3}He and ^{4}He, ^{6}Li and ^{7}Li due to a noticeable difference in the values of the thermal wavelengths of De Broglie (**De Broglie's Thermal Wavelength**) (see Fig. 55). The thermodynamic effects of this kind fade out with an increase in temperature, while isotopic effects in transport phenomena may not have a quantum nature.

$$\boxed{\text{J}}$$

Jamieson

John C. Jamieson was a crystallographer and physicist, an expert in high-pressure X-ray structural research, and a professor at the University of Chicago (Fig. 56). Together with A. Lawson, he began to

Fig. 56. John C. Jamieson (1924–1983) (from S. M. Stishov archive).

use diamonds for structural studies.[1] He proposed anvils with a pressurized boron gasket, which made it possible to obtain X-ray data at pressures of the order of 100 kbar.

Jamieson Award

The award was established by friends and students of Jamieson. The award is given to young scientists at the Gordon Conference (see **Gordon Conferences on High-Pressure**) and the AIRAPT Conference (see **AIRAPT**).

---

[1] Jamieson J. and Lawson A., Debye-Scherrer X-ray techniques for very high pressure studies, in *Modern Very High Pressure Techniques*, edited R. H. Wentorf, Jr., Washington: Butterworths (1962).

K

Kennedy

George C. Kennedy (1919–1980) — an American geophysicist — and geochemist, made a great contribution to the study of phase diagrams of simple substances, found maxima in the melting curves of cesium and tellurium, and built a cylinder–piston apparatus capable of generating pressures up to 100 kbar. He determined the pressures of reference points and created a new high-pressure scale.[1]

---

[1]Kennedy G. C. and LaMori P. N., *J. Geophys. Res.*, 67, 851 (1962).

$\boxed{\textbf{L}}$

Lasers and Dynamic Pressure

A short (nanosecond) laser pulse with a power of $>10^9$ watts/cm^2 focused on the surface of a solid target produces a plasma, the expansion of which, due to recoil, generates a shock wave with a pressure of $\sim 10^2$–10^3 GPa and a temperature of $\sim 10^4$K. In recent years, the dynamic compression method making use of a profiled laser pulse (ramp compression) has been developed (see **Dynamic Pressure**). This compression is not accompanied by a shock wave, which allows to obtain higher pressures of ~ 5 TPa. The procedure involves the use of many dozens of laser beams.[1]

Lentil

A lentil is a high-pressure cell developed in 1959 by the Institute of High-Pressure Physics of the Academy of Sciences of USSR.[2] The cell was named due to the shape of the working cavity, resembling a grain of lentil. This device made it possible to obtain the first artificial diamonds in Eastern Europe and was the basis for creating the industry of artificial diamonds in the USSR (see **Bridgman Anvils, Lentils, Toroid**).

---

[1]Smith R. F. *et al.*, *Nature*, 511, 330 (2014).
[2]Slesarev V. N., Vereshchagin L. F. and Ivanov V. E. (1962, unpublished).

Liquid Crystals

Liquid crystals are molecular systems whose molecules have cigar or disk-shaped form. Molecules of such systems can be ordered in space, thereby forming liquid crystals.[3] Liquid crystals are of great interest for the physics of phase transitions and are investigated at high pressures.

Liquid Media Transmitting Pressure

Any liquid inevitably solidifies at high pressures, so the choice of a pressure transmitting fluid to work in hydrostatic conditions has always been of particular importance. P. Bridgman worked with a 50% mixture of pentane and isopentane up to 30 kbar.

High grade gasoline also gives satisfactory results in this pressure range. It is argued that a mixture of methyl and ethyl alcohols, often used to create hydrostatic conditions in diamond anvils, does not harden at 100 kbar. At relatively low pressures, liquid oils, kerosene, glycerol, a mixture of glycerol with water, and silicone liquids are used. The latter are also used in the apparatus with a solid medium transmitting pressure. In this case, the test sample is in a special capsule filled with the appropriate fluid. In studies of neutron scattering, a fluorocarbon liquid (fluoroinert) that does not contain hydrogen, a strong neutron absorber, is used. For the same purpose, a deuterated mixture of methyl and ethyl alcohols is sometimes utilized. In recent years, Daphne oil (see **Daphne Oil**) became popular as a pressure transmitting medium.

Lithographic Stone

Lithographic stone is a limestone mixed with clay particles (mergel). It is found, like limestone, in the form of large bodies of sediments from the seas of the distant past. It is used as a pressure transmitting medium in cells like **Lentil** and **Toroid**.

---

[3]De Gennes P. G. and Prost J., *The Physics of Liquid Crystals*, Oxford: Oxford University Press (1993).

Lonsdaleite

Lonsdaleite is a hexagonal diamond, discovered in 1967 in the laboratory and then in the craters of impact origin in association with a regular diamond. It is named in honor of the British crystallographer Kathleen Lonsdale (1903–1971). She studied diamonds and discovered traces of nickel in artificial diamonds, thereby revealing the recipe for their synthesis. Lonsdaleite is found only in the form of traces, and there is no consensus about its identity. Many believe that lonsdaleite is nothing more than a diamond with a broken sequence of layers (stacking faults).

Low-Temperature Technique

The technique of low temperatures at high pressures has some specifics.[4,5] The high-pressure cells are rather bulky, and the refrigerant consumption during their cooling is very high. High-pressure cells (see **High-Pressure Enclosure**) designed for low-temperature studies can be distinguished into the autonomous clamp cells with fixed pressure and the cells connected to a pressure generator through a high-pressure tubing or a system of thin-walled tubes that transmit force from the press situated at room temperature. The clamp cells must be removed from the cryostat and reheated to change the pressure. The cells connected by tubing with sources of compressed gas or liquid may not be removed from the cryostat (sometimes this is important), but a cryostat should be warmed up. The enclosures using an external source of force require neither one nor the other, but their use is limited. An almost perfect example of a low-temperature research enclosure is the membrane diamond anvil (see **Diamond Anvils**), pressure in which is controlled by pressure of helium in the membrane, and can be changed at low temperatures.

---

[4]Swenson C. A., *Solid State Physics*, Vol. 11, Cambridge: Academic Press (1960).

[5]Stewart J. W., *Modern High-Pressure Techniques*, edited R. H. Wentorf, Jr. Washington: Butterworths (1962).

M

Manganin

Manganin is an alloy of copper ($\sim$85%), manganese ($\sim$12.5%) and nickel ($\sim$2.5%) with high electrical resistivity of $\sim$0.5$\cdot$10^{-6} Ohm$\cdot$m. The dependence of the electrical resistance of manganin on temperature passes through a mild maximum in the region of room temperature, where it has an extremely small temperature coefficient of resistance. This permits the use of manganin wire for the manufacture of reference resistance coils. One also can use an alloy of gold with chromium (2.1%), which has an even lower temperature coefficient of resistance than manganin.

Manganin Gauge

It was found many years ago that the electrical resistance of a manganin wire is almost linearly dependent on pressure. This property of the manganin made obvious the use of a manganin wire coil as a pressure gauge.

Typically, a manganin gauge is a frameless coil with a resistance of about 100 ohms, wound with a wire $\sim$0.05 mm thick in a silk insulation (Fig. 57(a)). A thicker wire (0.1–0.15 mm in diameter) in a varnish insulation can be also used for the manufacture of pressure gauges (Fig. 57(b)). After fabrication, the wire is annealed at temperatures of about 140°C for several days and subjected to aging at high pressures. Next, the manometers made must be calibrated, although

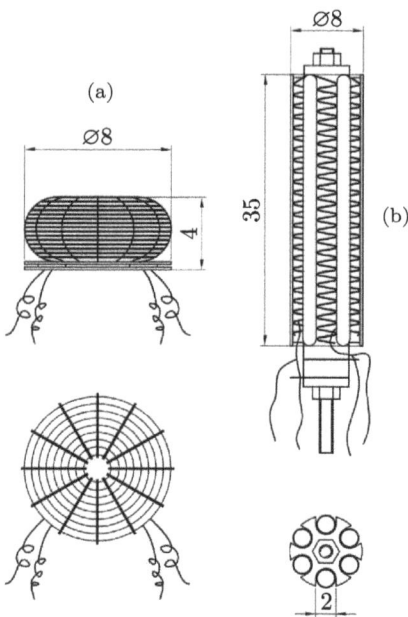

Fig. 57. Two types of manganin gauges: (a) loosely wound coil of manganium wire in a silk insulation with resistance of 100 ohms, (b) spiral of manganin wire in a varnish insulation with a resistance of 75 ohms on a ceramic frame.

the approximate value of the piezocoefficient of the manganin wire is well known as $\Delta R/R_0 = (2.3-2.5) \cdot 10^{-6}/\text{bar}$.

Calibration should be carried out using a reference Deadweight manometer (see **Deadweight Pressure Gauge**), although this is often done using reference points (for example, the crystallization pressure of mercury is 7,640 kg/cm^2 at a temperature of 0°C). The pressure dependence of the manganin resistance is weakly nonlinear, and this should be taken into account when processing the calibration results. As the manganin gauge operates, its resistance slightly increases, and although the piezo coefficient does not change much, the zero resistance value should be measured before each experimental cycle.

Marriott

Edme Mariotte (1620–1684) was a French physicist and one of the founders of the Paris Academy of Sciences. Independently from Boyle (see **Boyle**), he discovered the relationship between pressure and volume, known now as the Boyle–Mariotte law.

Maximum on the Melting Curve

Gustav Tammann (1861–1938) was German and Russian physical chemist who believed that the melting points of substances pass through a maximum on compression. However, P. Bridgman (see **Bridgman**), on the basis of his experimental studies, concluded that the melting curves of all substances increase indefinitely when pressure is applied. However, temperature maxima were found as the number of studied substances increased and high-pressure technology improved. Figure 58 shows the phase diagram of cesium, displaying distinct maxima on the melting curve. Temperature maxima were detected in a number of substances, however, it turned out that after the completion of a series of phase transitions, the melting curves acquire the shape predicted by Bridgman. We also point out that quantum effects can cause the appearance of temperature maxima at very high pressures.[1]

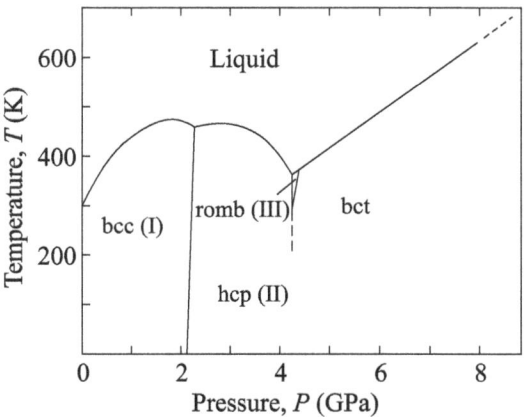

Fig. 58.   Cesium phase diagram.

---

[1]See details in Young D. A., *Phase Diagram of Elements*. Berkley, Los Angeles, Oxford: Univ. of California Press (1961).

Measurement of High-Pressures

Measurement of high-pressures is carried out with the help of special instruments called manometers. Among these devices are primary manometers and secondary manometers. Primary manometers are high-pressure apparatuses that allow absolute pressure values to be obtained, while secondary manometers need to be calibrated using primary ones. Secondary manometers are usually used during routine measurements. A Deadweight pressure gauge is used for absolute pressure measurements (see **Deadweight Pressure Gauge**). Bourdon manometers, manganin gauges and other secondary sensors are calibrated with the help of a Deadweight pressure gauge. However, the capabilities of the Deadweight gauge are limited to a pressure of 20 kbar. At higher pressures up to 30 kbar, extrapolation is used. In the development of an apparatus capable of generating pressures up to 100 kbar and above, it has become common to use reference points, such as abrupt changes in electrical resistance during phase transitions in some metals. The pressure in the experimental cell can be easily determined from the oil pressure in the press cylinder. The resistivity jumps in bismuth, thallium, barium, lead, etc. were taken as reference points. Initially, the pressure values of these transitions were drawn from P. Bridgman's data obtained in the anvils. Subsequently, J. Kennedy measured the pressure of the corresponding transitions using a piston piezometer[2] (see **Piezometer**), which led to a revision of the high-pressure scale. Currently, the following values of the most common fixed points are accepted:

Phase transition	Pressure, kbar
Bismuth (Bi) I-II	25.3
Bismuth (Bi) II-III	26.8
Thallium (Tl)	37
Barium (Ba)	55
Bismuth (Bi) V-VII	77
Barium (Ba)	123
Lead (Pb)	134

[2]Kennedy G. C. and LaMori P. N., *J. Geophys. Res.*, 67, 851 (1962).

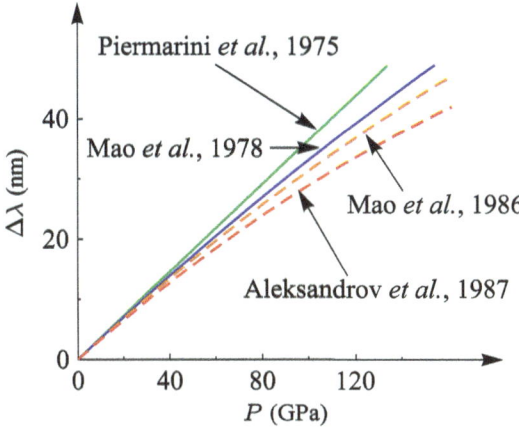

Fig. 59. Dependence of the shift of ruby luminescence line on pressure according to various data. The Alexandrov scale is constructed using *a priori* equation of state of diamond (see details in Alexandrov I. V. *et al.*, *JETP*, 93, 680 (1987)).

New problems with measuring high pressures arose with the invention of diamond anvils (see **Diamond Anvils**). These problems were successfully solved with the help of ruby $(Al_2O_3 + Cr)$, the luminescence of which serves as a means of measuring pressure. Initially, the pressure calibration of the ruby luminescence line was performed using the semi-empirical equation of state of NaCl, which resulted in a linear ruby scale (Piermarini, 1975, Fig. 59). This scale was corrected twice using shock data and X-ray studies of a number of metals (Mao 1978, 1986, Fig. 59). In 1987, a high-pressure diamond scale was proposed based on the equation of state of a diamond.[3] Later, the high-pressure scale for diamond anvils was repeatedly changed.[4]

Finally, we note that there is the possibility of constructing an absolute scale of high pressures using carrier substances. This feature, first noted by A. Ruoff, consists in measuring the bulk modulus of a carrier substance as a function of volume.[5] By integrating this dependence, it is possible to obtain the absolute pressure at each value of the carrier volume. An attempt to construct a high pressure

---

[3] Alexandrov I. V. *et al.*, *JETP*, 93, 680 (1987).

[4] For more details, see Syassen K., *High Pressure Res.*, 28, 75 (2008).

[5] Ruoff A. L. *et al.*, *J. Appl. Phys.*, 6, 1295 (1973).

scale for diamond anvils using MgO as the carrier (see **MgO — High-Pressure Scale**) was made in Ref.[6].

Measurement of Physical Quantities at High-Pressures

The development of, for example, multi-anvil apparatuses, diamond anvils, new materials, new sources of neutron and synchrotron radiation, permit for a variety of measurements at high pressures. These include thermodynamic values (heat capacity, thermal expansion, compressibility, magnetostriction), electric and magnetic quantities (electrical resistance, magnetoresistance, Hall effect, De Haas–Van Alven effect, Shubnikov–De Haas effect, magnetic moment), optics (absorption and reflection of light, Raman and Brillouin scattering), X-ray and neutron scattering (EXSAFS spectroscopy, Mossbauer effect, crystal structures and lattice dynamics, magnetic structures and small-angle scattering).

Melting

Melting is the transition of a solid crystalline substance to a liquid state. It occurs through a phase transition of the first order, and, accordingly, is accompanied by a jump in volume and entropy. The melting point generally grows without limit with pressure, although a different pattern can be observed locally, for example, the melting point of ice, silicon and germanium decreases with increasing pressure at low pressures, or passes through a maximum, as in the case of cesium (see **Maximum on the Melting Curve**).

Membrane Diamond Cell

In the membrane diamond anvil cell, the diamond anvils, driven by a membrane, are loaded with gas under controlled pressure. The advantage of this design is the possibility of changing the pressure

---

[6]Chang-sheng Zha *et al.*, *Earth. Planetary Sci. Lett.*, 159, 25 (1998).

without removing the cell from the devices and instruments intended for measurements (see **Diamond Anvils**).

MgO — High-Pressure Scale

MgO — high-pressure scale was constructed using experimental data characterizing the behavior of the bulk modulus of MgO as a function of volume (see **Measurement of High-Pressures**).

Mica

Micas are a group of layered minerals — aluminosilicates. In the high-pressure technique, muscovite $KAl_2[AISi_3O_{10}](OH)_2$ and phlogopite $KMg_3[AISi_3O_{10}](OH)_2$ are mainly used. Mica can be split into very thin leaves, suitable for the manufacture of various insulating gaskets, sleeves, etc. needed for electrical inputs in high-pressure apparatuses. Splitting mica is best done in water. Water wets mica, and penetrating between the layers, promotes the splitting.

Molecular Dynamics

Molecular dynamics, the method of determining numerical solutions of problems of statistical physics, consists in integrating the Newtonian equations of motion.[7] It was developed by Berni Alder of Livermore National Laboratory in the 1950s. The method is used to study the thermodynamic and kinetic properties of matter at ambient and high pressures.

Monte-Carlo Method

The Monte-Carlo method in statistical physics is designed to calculate the partition function by averaging the system states obtained

---

[7]Rapaport D. C., *The Art of Molecular Dynamics Simulation*, Cambridge: Cambridge University Press (1997).

as a result of random steps in the configuration space. It is used to study phase transitions and phase diagrams at high pressures.[8]

Multi-anvil Installations

Multi-anvil installations are characterized by a three-dimensional (spatial) arrangement of the working elements (pistons), as shown in Fig. 60. The specific design of the installations may be different (see, for example, the steel frame setup of the Hall (Fig. 43), the Split Spheres of Ruoff (Fig. 61) and Von Platen (Fig. 83)). Attention should be paid to Fig. 60(c), showing a set of truncated cubes of hard alloy or diamond compact, which can contain samples for research in the octahedral cavity. This system is placed in a cubic

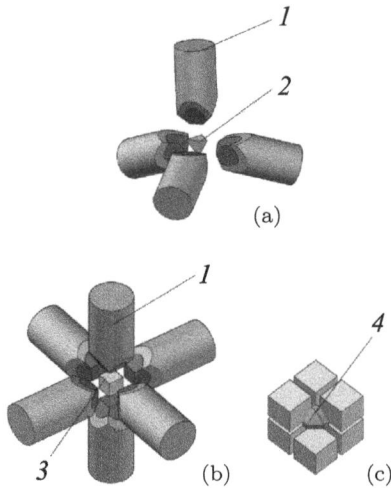

Fig. 60.   Spatial schemes of multi-anvil installations: (a) tetrahedral installation, (b) cubic installation, (c) second stage of a cubic installation, proposed by the Japanese scientist N. Kawai: 1 — piston, 2 — pyrophyllite tetrahedral, 3 — pyrophyllite cube, 4 — pyrophyllite octahedral, which is in a cavity formed by eight truncated hard alloy cubes.

---

[8]See more in Binder K. and Heermann D. W., *Monte-Carlo Simulation in Statistical Physics: An Introduction.* Berlin: Springer-Verlag (2010).

Fig. 61. Photograph of a split sphere forming a six-anvil system, built by A. Ruoff in 1977 in the hope of producing metallic hydrogen (from S. M. Stishov archive).

installation instead of a pyrophyllite cube (3). With this arrangement, the pressure can reach several dozen GPa.

Mushroom Seal

The mushroom seal is a type of high-pressure seal based on the principle of unsupported area (for details, see **Seal**).

$$\boxed{\text{N}}$$

Nanodiamonds

Nanodiamonds are diamonds smaller than 10 nm (nanometers) that are produced by explosions, meteorite impacts, decomposition of hydrocarbons under pressure, CVD, etc. With appropriate doping, nanodiamonds can be used in electronics, medicine, quantum computing and in other applications.

Nitrogen

Nitrogen, N_2, is a gas, an integral part of air. The critical temperature of nitrogen is 126.2 K. It condenses at 77.4 K and crystallizes at 63.29 K. Liquid nitrogen is widely used as a refrigerant. At room temperature, it crystallizes at a pressure of ~28 kbar. It is used as a pressure transmitting medium. At very high pressures, attained in diamond anvils, the triple bonds in the N_2 molecule break, and the nitrogen enters an atomic state with the single interatomic bonds.[1]

---

[1] Eremets M. I., Gavriliuk A. G., Trojan I. A., Dzivenko D. A. and Boehler R., *Nat. Mater.*, 3, 558 (2004).

O

O-ring

O-rings are widely used in industry for movable and stationary seals. The O-shaped seal is a torus of circular cross-section made from oil-resistant rubber or other sufficiently elastic material, for example, from Viton (Viton–fluorocarbon rubber) (Fig. 62(a)). Figures 62(b_1) and 62(b_2) show two ways to install an O-ring for a seal in a piston system. The principle of action of the O-ring is illustrated in Fig. 62(c_1)–(c_3). The width of the groove for installing the O-ring should be somewhat smaller than the diameter of the ring (prescribed in industrial standards), and there is some overpressure that prevents leakage of liquid or gas.

When the medium pressure increases, the ring closes the gap between the elements of the installation (Fig. 62(c_1)). With a further increase of pressure, the ring is squeezed into the gap and collapses (Fig. 62(c_2)). This situation is corrected by means of a protective ring (Fig. 62)c_3)). Currently, rubber sealing O-rings are also used in high-pressure laboratory installations, replacing the seals proposed by Bridgman, due to significantly lower friction losses.

Obturator

The obturator is a plug or a closure. In the high-pressure apparatuses, the obturator is a fixed device that closes the high-pressure channel. The obturator often contains insulated electrical leads, optical windows, etc.

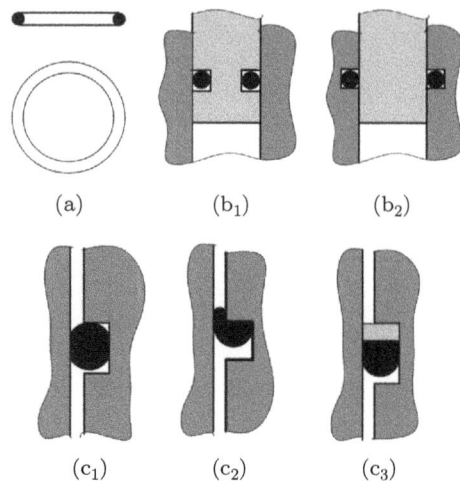

Fig. 62.   O-ring for seals in high-pressure apparatuses.

Olivine

Olivine is magnesium iron orthosilicate of composition $(Mg,Fe)_2SiO_4$. Along with pyroxene $(Mg,Fe)SiO_3$, olivine is the predominant mineral in the Earth's upper mantle. The final terms in the olivine series of solid solutions are called forsterite Mg_2SiO_4 and fayalite Fe_2SiO_4. The crystal structure of olivine can be described as the hexagonal closest packing of oxygen atoms with half of the octahedral holes occupied by Mg and Fe and the eighth part of the tetrahedral holes filled with Si.

Olivine–Spinel Transition

John Bernal (see **Bernal**) suggested for explanation of geophysical observations that olivine (see **Olivine**) in the depths of the Earth will acquire a spinel structure with increasing density. Such a transition actually exists at high pressures and corresponds to the transformation of the hexagonal packing of oxygen atoms into cubic, but the character of the distribution of cations in the holes does not change. The latter means that silicon is still in four-fold coordination (see also **Ringwood**).

$$\boxed{\textbf{P}}$$

Paronite

Paronite is a layered material made from a mixture of asbestos, rubber, etc. It is used in sealing gaskets.

Perovskite

Perovskite is a mineral having the composition $CaTiO_3$. The crystal structure of perovskite can be described as the close packing of oxygens with the inclusion of 1/4 of calcium atoms. Titanium atoms occupy 1/4 octahedral holes created by oxygens. The structural type of perovskite turned out to be very common among compounds of type ABX_3. It is the perovskite structure that pyroxene of the composition $(Mg,Fe)SiO_3$ acquires with silicon in six-fold coordination at high pressures. The natural analogue of this phase is called Bridgmanite (see **Bridgmanite**). It is believed that the perovskite composition $(Mg, Fe) SiO_3$ composes about 90% of the lower mantle of the Earth (see **Earth's Mantle**).

Phase Diagram

A phase diagram is a graphical display of the phase state of a substance as a function of external parameters: temperature, pressure, volume, magnetic and electric fields, etc. (see **Phase Transitions, Pressure in Nature**).

Phase Transitions

Phase transitions are transitions between different physical states (phases) of the same substance. Common examples of phase transitions include melting ice and boiling water, or transformation of graphite to diamond at high pressures. Figure 63 shows schematically the phase diagrams of a substance in the coordinates temperature–pressure $(T-P)$ and temperature–volume $(T-V)$.

Obviously, these simplified schemes do not reflect the entire diversity of phase diagrams and phase transitions. For example, the ice melting curve actually has a negative slope $(dT/dP < 0)$, while in Fig. 63 the melting curve is depicted with a positive slope. Moreover, the melting curves can have a maximum or, due to quantum effects, terminate at an absolute zero temperature. The solid region (Fig. 63) may include numerous phase transitions associated with changes in their crystal and electronic structure (see **Diamond**, **Pressure in Nature**, **Isostructural Phase Transitions**, **Iron**, **Stishovite**).

In the examples given, the phase transitions are accompanied by abrupt changes in the specific volume and entropy. Such transitions are called first-order phase transitions. Phase transitions of the first order mainly occur with a change of the aggregate state of a substance and radical changes in the crystal structure of solids.

Figure 64 illustrates the behavior of the Gibbs potential G during a phase transition of the first order. As can be seen, the temperature and pressure of the phase transition are determined by an intersection of the potential branches characterizing the coexisting phases

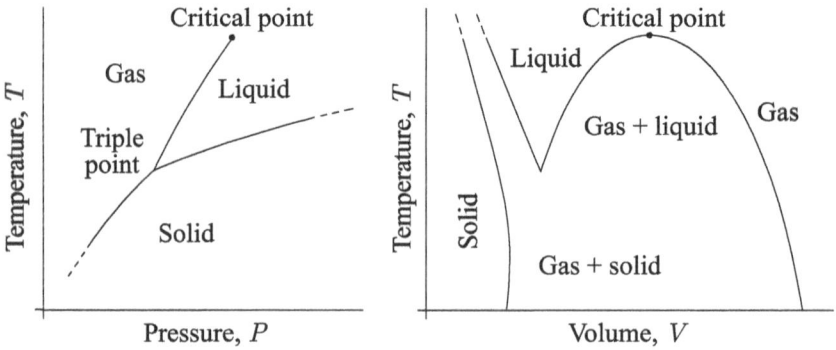

Fig. 63.   Phase diagrams of one-component system.

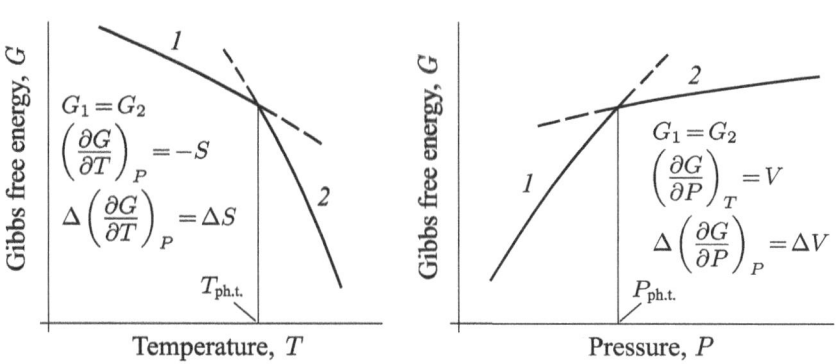

Fig. 64. Behavior of the Gibbs thermodynamic potential at a first-order phase transition.

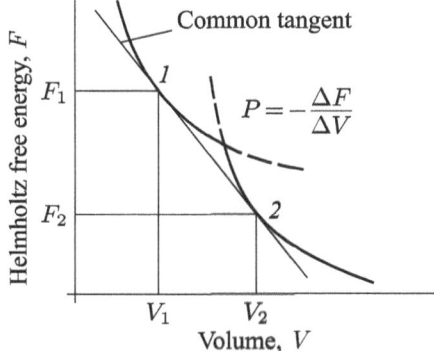

Fig. 65. Behavior of the Helmholtz free energy at a first-order phase transition.

of substance. The entropy and volume jumps follow from the very fact of the intersection of the thermodynamic potential branches. Pay attention to the fundamental existence of metastable states of a substance corresponding to the segments of the potential indicated by dashed lines. In some cases, it is convenient to consider a phase transition using the Helmholtz free energy F and the volume V as an independent variable. Then, as it follows from Fig. 65, the parameters of a first-order phase transition are determined by the common tangent to the free energy branches, and not by their intersection, as is the case for the thermodynamic potential G. Recall that the common tangent condition means equality of pressures of the coexisting phases.

Along with phase transitions of the first order, there is also an extensive group of phase transitions of the second order or continuous phase transitions, characterized by continuous changes in specific volume and entropy, but accompanied by peculiarities of the behavior of heat capacity, coefficient of thermal expansion, compressibility, etc. Transitions of the second order include transitions associated with the emergence of magnetism, superconductivity, superfluidity, orientational order, etc.

This classification, separated phase transitions as first and second order, proposed at the time by P. Ehrenfest, is based on orders of derivatives of the thermodynamic potential experiencing jumps during a phase transition. P. Ehrenfest suggested equations relating the slope of a curve of a second-order phase transition with jumps in heat capacity, compressibility, and thermal expansion coefficient. Subsequently, it was found that, for the most part, during phase transitions of the second order, divergence of the corresponding quantities is observed as a result of fluctuation effects.

Piezometer

A piezometer is a device for measuring the volume of a substance at high pressures. The simplest form of a piezometer is a cylinder–piston system, equipped with a piston displacement indicator and a gauge to measure the force applied to the piston (Fig. 66(a)). In the experiment, the displacement of the piston as a function of the load is measured. Friction in the seal of the press, friction in the seal of the piston and friction of substances against the cylinder wall, elastic hysteresis (see **Hysteresis**) of the loaded piezometer parts create the effect shown in Fig. 67.

Usually, when processing measurement results, it is assumed that the hysteresis loop is symmetric, which is apparently not quite correct. Additionally, it is required to make corrections for the deformation of the piezometer elements. Figure 66(b) illustrates a piston piezometer adapted to the variation in measurement temperature. Here, the piston system and the sample to be measured are separated in space and located at different temperatures. Naturally, in this case, when measuring the compressibility of solids, a medium (liquid) transferring pressure is used, which requires corrections

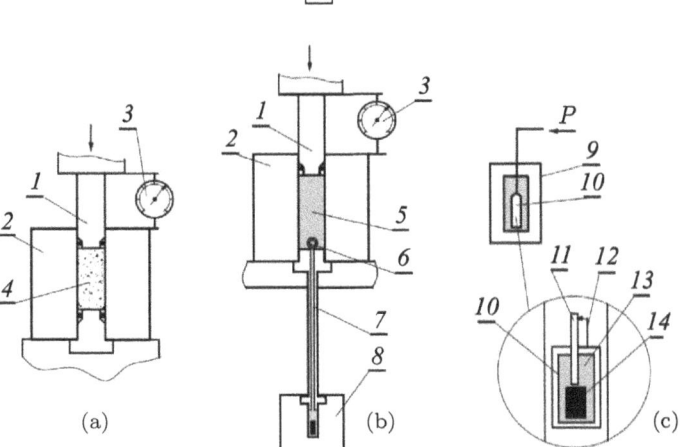

Fig. 66. Piezometers for measuring the compressibility of substances at high pressures: (a) simplest piston piezometer, (b) piston piezometer with an additional vessel, (c) immersible piezometer: 1 — piston, 2 — body, 3 — displacement gauge, 4 — sample, 5 — pressure transmitting fluid, 6 — pressure sensor, 7 — high pressure tubing, 8 — high-pressure vessel containing sample, 9 — high-pressure vessel, 10 — submersible piezometer, 11 — piston, 12 — displacement gauge, 13 — reference fluid, 14 — sample (see text).

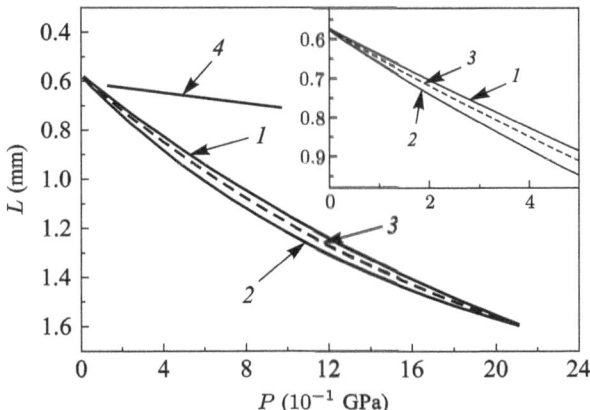

Fig. 67. The dependence of the length of the sodium sample on the nominal pressure according to measurements in the piezometer type (a) (Fig. 66): 1 — forward stroke, 2 — reverse stroke, 3 — "true" length, 4 — correction for system deformation (Swenson C. A., *Solid State Physics*, Vol. 11. Academic Press (1960)).

for the liquid compressibility and internal volume change of the pressure apparatus. Figure 66(c) shows a piezometer directly in a pressure-transmitting fluid. Here, the piezometer is under hydrostatic pressure, which simplifies the introduction of deformation corrections. When measuring solids in an immersed piezometer, a reference fluid is used, the equation of state of which is well known. A piezometer — dilatometer, for measurements in purely hydrostatic conditions, is shown in the subsection **Dilatometer**.

Piston

A piston is a cylindrical piece made of a hard material (steel, hard alloy) used in high-pressure apparatuses to transfer pressure to an object or a medium in a cylindrical cavity (see **Cylinder–Piston**).

Press Hydraulic

A hydraulic press is a power device that uses fluid pressure to create a force acting on an object. An example of the design of laboratory press is shown in Fig. 68. Here, 1 and 2 are the elements of the power cylinder, 3 are plates, 4 columns, 5 thrust block, 6 nuts, 7 is the inlet for hydraulic fluid that controls the forward stroke of the piston and 8 is the inlet for the reverse-acting fluid. The generated force of the press is $F = pS$, p — pressure, S — the area of the piston. The optimal pressure of fluid in the cylinder for the case of strong structural steel with a yield strength of $\sim$90 kg/mm^2 amounts to 1,500–2,000 kg/cm^2, which actually determines the dimensions of the cylinder and the press at the required press power.

Figure 69 shows a photograph of a 1,000-ton double-cylinder press. As a rule, capacity of 1,000–2,000 tons is sufficient for most laboratory studies, although several laboratories in the world have presses with a capacity of 30,000–50,000 tons. In the case of presses of such high power, special designs of the frames are used.

In particular, a powerful press can be built in the form of pipes with an operating window, plugged at both ends with a working cylinder and a base plate. In other cases, the power frame may be a structure, bonded with tension-wound steel tape. The press must be equipped with a valve system (see Fig. 46), allowing the forward and

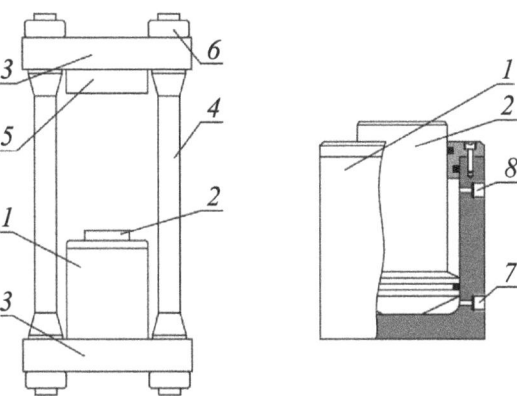

Fig. 68. Scheme of the hydraulic press. Power cylinder of the press is shown separately on the right.

Fig. 69. 1,000-ton two-cylinder hydraulic press.

reverse stroke of the piston. However, in the case of small laboratory presses with capacity of up to 20 tons, it is worthwhile to use a powerful spring to realize the piston reverse stroke (see Fig. 70).

Pressure — Definition and Units

The mechanical definition of pressure is a force applied to an area $P = F/S$, which, however, is equivalent to the thermodynamic determination of pressure as a derivative of free energy over the volume

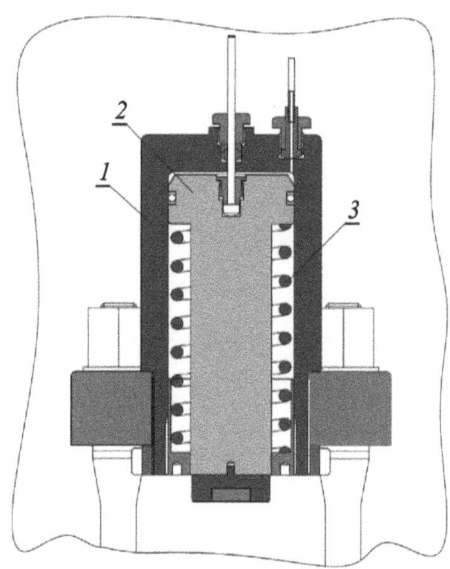

Fig. 70. Cylinder of a laboratory hydraulic press with a return spring: 1 —
cylinder body, 2 — piston, 3 — spring.

$P = -(\partial F/\partial V)_T$. Pressure is measured in Pascals (Pa) in the SI system and bars (Bar) in the CGS system. In industry, the unit kg/cm^2, called a technical atmosphere, is often used. The conversions for the units are

$$1\,\mathrm{Pa}\,(1\,\mathrm{N/m^2}) = 10\,\mathrm{dyn/cm^2} = 10^{-5}\,\mathrm{bar} = 1.02\cdot 10^{-5}\,\mathrm{kg/cm^2}.$$

In some cases, it is useful to introduce natural pressure units. For example, in case of a substance whose energy is determined by two quantities: the characteristic energy ε and the characteristic length σ, a natural pressure unit, based on the dimension, is $P_0 = \varepsilon/\sigma^3$. We point out that for helium He — $P_0 = 84$ bar, for xenon Xe — $P_0 = 450$ bar. These values can be used to build reduced equations of state, reduced phases of diagrams and make various kinds of predictions. At the atomic level, the atomic unit of pressure is $e^2/a_0^4 \approx 3\cdot 10^4$ GPa, here $e^2/a = 2\mathrm{Ry} \approx 27.2$ eV, $a_0 = h^2/me^2$ is the Bohr radius. This value roughly corresponds to the ionization pressure of the substance (transition to the metallic state). The pressure of the total ionization of a substance is an order of magnitude $Z^5 e^2/a_0^4 \sim Z^5 \cdot 10^4$ GPa, where Z is atomic charge (see **Pressure in Nature**).

Pressure in Nature

The following table shows the pressure and density values in the Earth and a number of space bodies:

	Pressure (Mbar)	Density (g/cm^3)
Earth (center)	4	~10–20
Exoplanets of Earth group (center)	>10^4	>20
Sun (center)	10^5	10^2
White dwarfs	10^{10}	10^6
Neutron stars	10^{22}	10^{14}

The evolution of a substance with increasing pressure can be described using the following sequence (see also Fig. 71):

Gas $\Leftrightarrow$ Liquid $\Leftrightarrow$ Crystal ... Crystal $\Leftrightarrow$ Metal $(e^2/a_0^4 \approx 3 \cdot 10^4 \, \text{GPa})$ $\Leftrightarrow$ Fully ionized substance $(Z^5 e^2/a_0^4 \sim Z^5 \cdot 10^4 \, \text{GPa})$ $\Leftrightarrow$ White dwarfs) $\Leftrightarrow$ Nuclear reactions (Neutron stars) $\Leftrightarrow$ Quark stars....[1]

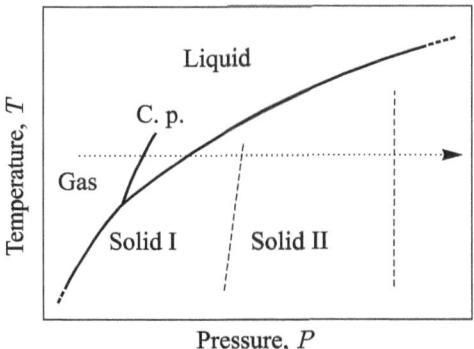

Fig. 71.   Scheme illustrating the behavior of a substance under compression along the dotted line.

---

[1]See details in Kirzhnits D. A., *UFN*, 104, 489 (1971).

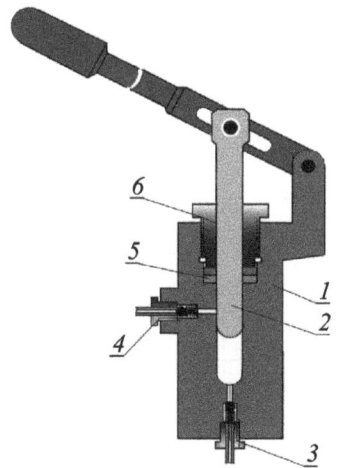

Fig. 72. High-pressure manual pump: 1 — body, 2 — plunger, 3, 4 — inlet and suction valves, 5 — sealing gland, 6 — base bushing.

Pump

A high-pressure pump produces compressed fluid intended to control hydraulic devices: hydraulic presses, intensifiers, valves, etc. A simple version of a laboratory hand pump is shown in Fig. 72.

A manual pump of this type can compress liquid up to pressures of $\sim$2 kbar. Mechanically driven laboratory pumps contain a similar plunger system with valves and, therefore, are not fundamentally different from a hand pump.

Pyrophyllite

Pyrophyllite is a layered mineral of the composition $Al_2[Si_4 O_{10}](OH)_2$. It is soft, easy to handle. After firing up to $1,000°C$, it loses water and turns into a kind of ceramic. It is used in high-pressure experiments as a pressure-transmitting medium in multi-anvil apparatuses (see **Multi-anvil Installations**).

Talc, a mineral of $Mg_3[Si_4O_{10}](OH)_2$, belongs to the same mineral class. Talc is significantly softer than pyrophyllite and is not used in the technique in its raw unbaked form. The ceramic made from talc powder is called steatite.

$$\boxed{\text{Q}}$$

Quantum Critical Point

The quantum critical point corresponds to the vanishing of the long-range order in case of a second-order phase transition at $T = 0$ K (see **Quantum Phase Transitions**).

Quantum Phase Transitions

Quantum phase transitions are transitions of the continuous type, occurring at $T = 0\,\text{K}$ as a result of quantum fluctuations, the existence of which is determined by the Heisenberg uncertainty principle. Quantum phase transitions occur at a certain critical value of the control parameter (pressure, magnetic field, concentration) governing the intensity of quantum fluctuations.[1]

Quartz

Quartz, a mineral of composition SiO_2, is widely distributed in nature. It has two polymorphic modifications: α and β. When heated, it transforms into polymorphic phases: tridymite and cristobalite (see **Silica**).

---

[1]Stishov S. M., *Phase Transitions for Beginners*. Singapore: World Scientific (2018).

Quenching

Quenching, in the heat treatment of metals and alloys, is a rapid cooling of the heated part in order to maintain a favorable crystal structure and/or texture of the material (see **Heat Treatment**).

R

Receiver

A receiver is a high-pressure vessel that acts as a kind of accumulator that stores a volume of compressed liquid or gas, sufficient for subsequent powering of hydraulic or gas devices. The required amount of compressed fluid is supplied to certain mechanisms through a fine-tuning valve, which avoids pressure pulsations that are inevitable when directly supplied from a pump or compressor.

Rhenium

Rhenium is a rare transition metal. It is chemically inert and easily hardened during plastic deformation. Rhenium is widely used for the manufacture of gaskets utilized in diamond anvils.

Ringwood

Alfred Edward Ringwood was an Australian geophysicist and geochemist (Fig. 73). He graduated from the University of Melbourne (Australia) with honors in 1953. In 1956, he received the degree of Doctor of Philosophy (PhD) under the leadership of A. Gaskin and F. Birch. Ringwood was fascinated by the Bernal idea of the role of the olivine–spinel transition in the formation of the structure of the Earth's depths. Investigating the germanium analogs of olivines and pyroxenes and their solid solutions with silicates, Ringwood was able to predict the fate of the most common minerals in the Earth's upper mantle. In 1966, he demonstrated that Fayalite Fe_2SiO_4,[1] the final member of the continuous olivine series (see Olivine), really takes on a spinel structure at high pressures. Later, this transition was confirmed for the case of a solid solution $(Mg,Fe)_2SiO_4$,[2] with a content of the magnesium component of up to 85%.

Fig. 73. A. E. Ringwood (1930–1993) (from S. M. Stishov archive).

Ringwoodite

In 1969, in the Tenham meteorite, a mineral $(Mg,Fe)_2SiO_4$ was found with a spinel structure which was named "ringwoodite" in honor of A. E. Ringwood (see **Ringwood**), who put a lot of effort to prove the existence of the dense phase of magnesium-ferrous orthosilicate.[3]

Ruby Gauge

The ruby manometer is a secondary pressure sensor based on the shift of the ruby luminescence line with pressure (Fig. 74). Measurement accuracy is low and does not exceed 300–500 bar. It is mainly used to measure very high pressures generated in diamond anvils (see **Measurement of High-Pressures**).

[1]Ringwood A. E., *Am. Mineral.*, 44, 659 (1959).

[2]Ringwood A. E. and Major A., *Earth Planet. Sci. Lett.*, 1, 241 (1966).

[3]Binns R. A., Davis R., Reed J. S. and Stephen J. B., *Nature*, 221, 943 (1969).

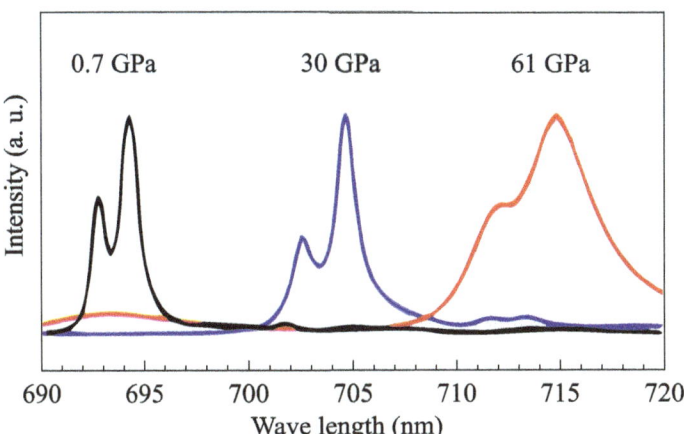

Fig. 74. R1–R2 ruby luminescence lines as a function of pressure in compressed helium. Distortion of the form of the doublet indicates the occurrence of non-hydrostatic stresses at high pressures.

$$\boxed{\text{S}}$$

Safety

Some high pressure apparatuses may contain stored mechanical strain energy and, depending on the working pressure fluid, a large amount of fluid compression energy. A catastrophic failure can produce the following high-pressure hazards: (1) a shock wave, (2) a fluid jet and (3) high speed solid fragments. Potentially dangerous parts of the apparatus must be shielded.[1] The founder of high-pressure physics, P. W. Bridgman, clearly understood the risks connected with high-pressure experimentation. They say that he taught his students: "Never take out a stuck piston using your right hand!"

Salt (NaCl)

NaCl is often used as a material that isolates a sample and as a quasi-hydrostatic pressure-transmitting medium in a "solid-phase" high-pressure apparatus.

---

[1]Sherman W. F. and Stadtmuller A. A., *Experimental Techniques in High-Pressure Research*, New York: John Wiley and Sons Ltd (1987).

Seal

The movable and stationary seals in high-pressure apparatuses can be divided into two groups:

(1) Seals operating as a result of pre-application of force so pressure in the seal always exceeds pressure of the medium.
(2) Seals operating on the principle of unsupported area proposed by P. Bridgman, when pressure in the seal automatically exceeds pressure of the medium by a certain percent determined by the geometry of the seal (see Fig. 75).

Seals (1) are used in low-pressure devices: pumps, compressors, valves. With reasonable sizes of various devices, it is not practical and possible to create a pre-pressure of more than 5 kbar in the seal. Though the "cone-in-seat" or "sphere-in-seat" seals for connections of high-pressure tubings can be used at least to 15 kbar (see **Elements of High-Pressure Installations**).

Seals (2) operate in a wide pressure range up to 30 kbar and beyond. Problem with the seal (a, b, d) is a "pinch off" of the

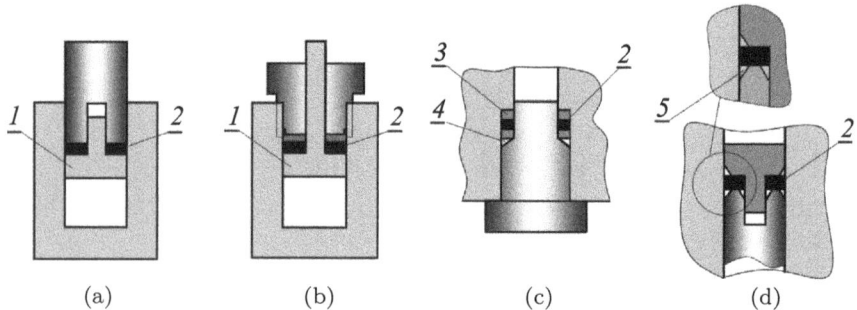

(a)   (b)   (c)   (d)

Fig. 75.   Movable (a) and stationary (b) seals of the unsupported area type (type I after Bridgman), unmovable (c) seal of the unsupported area type (type II after Bridgman), movable seal (d), equipped with anti-extrusion rings: 1 — mushroom like seal, leg of which has no support, enables excessive pressure in the gasket 2, 3 — metal durable rings, 4 — empty space of triangular cross-section, creating the effect of uncompensated area, 5 — anti-extrusion rings.

mushroom leg and a great friction for a moving piston. Currently, however, more O-ring-based seals are used (see **O-ring**).

Separator

A separator is a device for separating two different media, for example, gas or solution and oil under high pressure. The need for such a device arises when direct compression of one of the media with a compressor or pump is undesirable or impossible. In the simplest case, the separator is a cylinder with a movable piston separating one cavity from another. Sometimes mercury in a U-shaped tube, a rubber or metal membrane is used as a movable piston. The intensifier (see **Intensifier**), when compressing the experimental gas, using the oil pressure in the low-pressure cylinder, can also be considered as a separator. A membrane compressor (**Compressor Membranic**) is also an example of a separator.

Silica

Silica is a silicon–oxygen compound of composition SiO_2. It is found in the form of numerous crystalline modifications and amorphous phases (agate, opal, chalcedony, jasper). The behavior of silica at high pressures largely determines the structure and mineral composition of the lower mantle of the Earth, the terrestrial planets and, possibly, exoplanets. Figures 76 and 77 show phase diagrams of silica at high pressures.

Note that the phase transition in SiO_2 from the α-PbO_2 to the pyrite structural type FeS_2 means an increase in the coordination number of silicon from 6 to 6 + 2.[2] The pressure of the silica phase transition to the pyrite structure is much higher than the pressure at the boundary of the Earth's mantle and, therefore, this transition cannot be related to the processes occurring in the

---

[2]Kuwayama Y. *et al.*, *Science*, 309, 923 (2005).

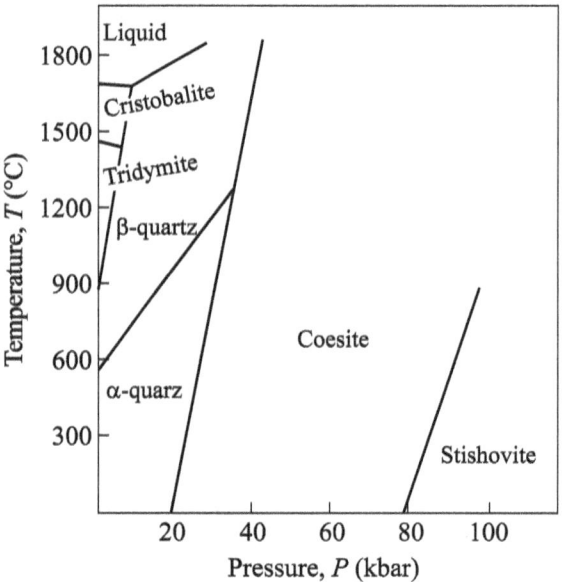

Fig. 76.   Phase diagram of silica SiO_2.

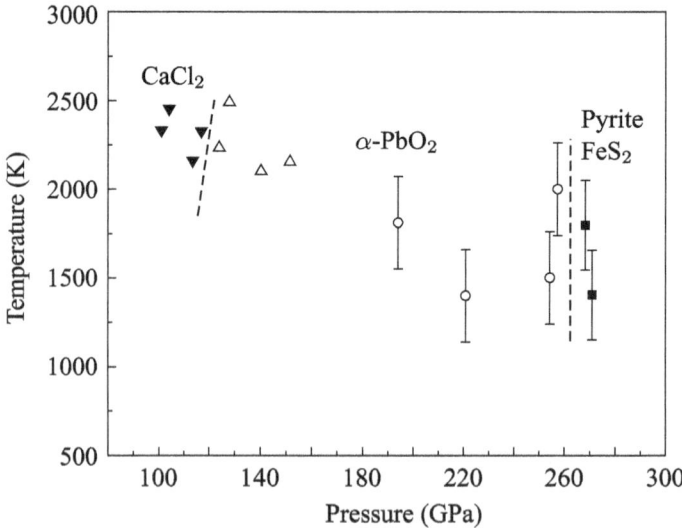

Fig. 77.   Silica phases at pressures above 100 kbar (Chao E. C. T., Fahey J. J., Littler J. and Milton D. J., *J. Geophys. Res.*, 67, 419 (1962)).

depths of the Earth. However, it is possible that the "pyrite" form of silica exists in the mantles of the exoplanets.

Silver Chloride (AgCl)

AgCl is a soft plastic substance, easily obtained by precipitation from a solution of silver nitrate. It is used as a quasi-hydrostatic medium when calibrating solid-phase apparatuses using the electrical resistance of reference metals.

Steel and Alloys

Steel and alloys used in mechanical engineering may be divided into two categories: structural and tool. Structural steels have high tensile strength, bending and torsion, while tool steels are required to have high compressive strength and hardness. The usual strength of structural steels with a hardness of HRc 40–50 after heat treatment is $\sim$150–200 kg/mm^2 with an elongation at break >5%. The compressive strength of tool steels with a hardness of $\sim$60 HRc is characterized by a value of 250–300 kg/mm^2. Accordingly, structural steels are used to manufacture parts and elements working in tension and bending: high-pressure vessels, press cylinders, columns, etc. Tool steels are used in the manufacture of pistons and other parts working in compression.

One should also be aware that there is a class of carbon-free, highly alloyed steel-alloys with high strength and ductility (maraging steels). These steels belong to the so-called precipitation hardening steels class and gain strength in the process of aging.

Beryllium bronze (see **Beryllium Bronze**) is another important material used in high-pressure engineering. Aluminum alloys are also used in high-pressure for special purposes (see **Windows for Hard Radiation**).

Finally, hard alloys — tungsten carbide, cemented with cobalt (WC + 2–20% Co) — are an unsurpassed material for the manufacture of pistons and anvils. The compressive strength of individual grades of cemented carbide exceeds 600 kg/mm^2.

Stishovite

Stishovite is a mineral of composition SiO_2 with a rutile-type crystal structure (TiO_2), where silicon is surrounded by six oxygen atoms. First discovered in the Arizona meteorite crater in 1961 (see **Chao, Crater**).[3] Subsequently it was found in almost all large structures (craters) of impact origin (Vredfort Dome, South Africa; Sudbury, Canada). The inclusions of stishovite in diamonds that have risen from great depths have been described.[4] However, the stishovite natural finds were preceded by the discovery of a dense modification of silica with rutile structure in the laboratory[5] (Figs. 78 and 79).

The search for a dense modification of silica was motivated by the ideas of F. Birch (see **Birch**), who suggested that in the transition layer of the Earth's mantle (see **Earth's Mantle**), ordinary surface rocks and minerals should be transformed into dense phases

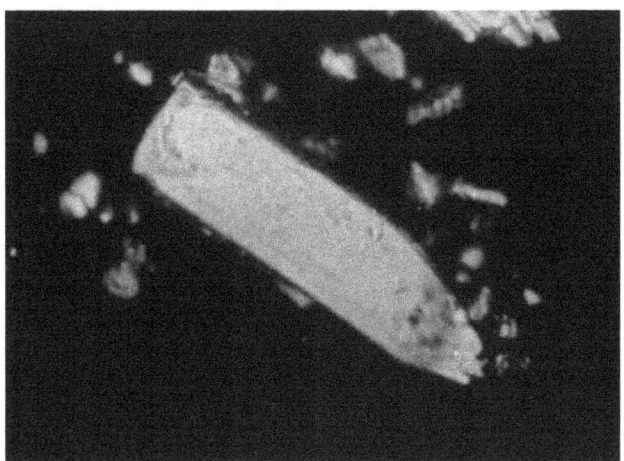

Fig. 78. Stishovite microcrystal (micrograph from Stishov S. M. and Popova S. V., *Geochemistry*, 10, 837 (1961)).

---

[3]Chao E. C. T., Fahey J. J., Littler J. and Milton D. J., *J. Geophys. Res.*, 67, 419 (1962).

[4]Kaminsky F., *Earth-Sci. Rev.*, 110, 127 (2012).

[5]Stishov S. M. and Popova S. V., *Geochemistry*, 10, 837 (1961).

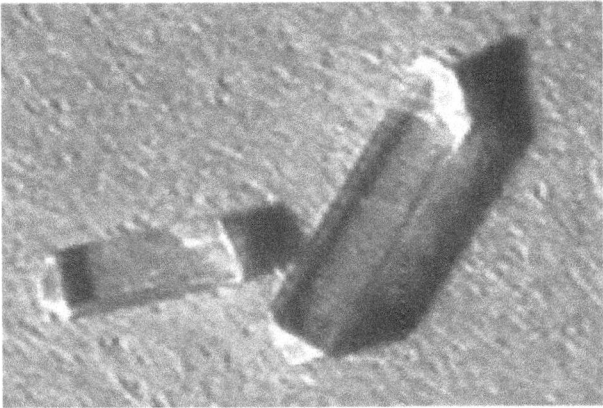

Fig. 79.   Stishovite single crystals up to 4 mm in size, grown in the HPPI by
N. A. Bendeliani.

with silicon in the six-fold coordination, which should significantly
increase their density and elasticity. Indeed, the density of stishovite
turned out to be 64 % higher than the density of quartz, while the
bulk modulus of stishovite exceeds that of quartz by almost an
order of magnitude (see the phase diagram of SiO_2 in the subsection **Silica**).

Strength

Strength is a concept characterizing the stability of a material in relation to its destruction or irreversible deformation (see
Deformation).

Strength Calculations of the Elements of High-Pressure Devices

A number of parts of high-pressure installations can experience significant stresses. Therefore, when designing it is necessary to carry
out the appropriate strength calculations.[6] Sometimes it's easy.

---

[6]Tsiklis D. S., *Handbook of Techniques in High-Pressure Research and Engineering*, New York: Plenum (1968).

For example, in the case of parts experiencing simple stretching or compression (columns of presses, various pistons), reference data characterizing the yield strength of the material under tension or compression can be used. Further calculations should take into account a margin of safety, which generally should not be less than two. When calculating the strength of thick-walled vessels, the situation is not so simple, and the appropriate theory should be used that links the strength of the vessel with standard reference data. Here, it must be borne in mind that with the strength of standard steel of about 150–200 kg/cm^2, a thick-walled single-layer cylinder, regardless of the diameter's ratio, will inevitably experience plastic deformation at pressures above ~5 kbar. To design devices for operation at higher pressures, this plastic deformation should be taken into account. The efficiency of vessels can be improved by making them multi-layered, which is also amenable to calculations (see **Support of Pressure Vessels**). Thread should be calculated for distortion and shearing, and should work only in the elastic regime; otherwise the threaded component cannot be disassembled. In loaded parts, the square and acme threads should be avoided. Finally, it seems that the best way to design high-pressure installations is to use the finite elements method software package, which allows for the identification of all hazardous stresses.

Structural Phase Transitions

Structural phase transitions are phase transitions associated with changes in the crystal structure. The order parameter characterizing the structural phase transition, in contrast to magnetic and super-conducting transitions, is determined by the features of the crystal structure. Thus, L. Landau, creating a theory of phase transitions of the second order, chose the value of the Ba atom displacement from the center of the elementary cell of barium titanate $(BaTiO_3)$[7] as the order parameter.

---

[7]Landau L. D. and Lifshitz E. M., *Statistical Physics*, Vol. 5 (3rd edn.), Oxford: Butterworth-Heinemann (1980).

Support of Pressure Vessels

To increase the maximum operating pressure of high-pressure vessels, pre-loading of the vessel with compressive "pressure" is used. One of the amplification methods is described in the **Autofrettage** subsection.

Another common way to strengthen vessels is to create multi-layer structures by hot press fitting or just press fitting external components to produce a residual compressive stress at the inner surface of the vessel (Figs. 80(a) and 80(b)). The method of variable mechanical support proposed by P. Bridgman is also used when the conical vessel moves into the support holder as the internal pressure increases (Fig. 80(c)). Another popular method of supporting vessels is to wind a strained steel strip in tension around the outside of the vessel (see **Von Platen**). In Fig. 80(a), one can see the outer

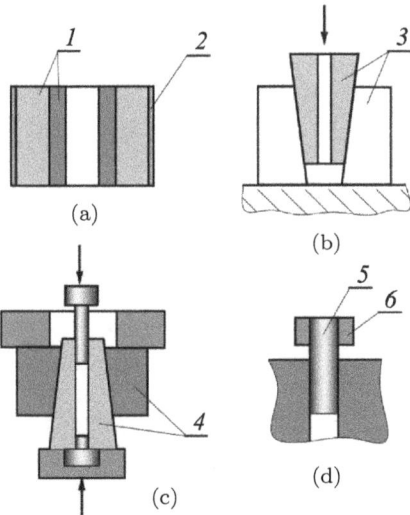

Fig. 80. Support for vessels and pistons of high-pressure apparatuses: (a) a multi-layer vessel made by means of a hot press-fitting process, 1 — steel rings made of durable, hardened steel, 2 — shell of mild steel, preventing fragments from scattering when the vessel is destroyed; (b) 3 — making a multi-layer vessel using the pressing method, creating a constant supporting force on the inner vessel; (c) 4 — creating a variable support of the vessel as the internal pressure increases; (d) supporting the piston face (5) using a hot-pressed steel ring (6).

thin shell, a "safety ring". This shell is made of soft metal. The purpose of the shell is to reduce scattering of the fragments of the burst vessel. It should also be noted that the loaded ends of the pistons are often prone to cracking. To avoid this situation, it is useful to cover the end of the piston with a steel ring set on a press fit, as shown in Fig. 80(d).

$$\boxed{\text{T}}$$

Talc

See the subsection **Pyrophyllite**.

Teflon

Teflon, polytetrafluoroethylene, is a plastic, extremely inert and heat-resistant material with very low friction, that can be easily machined. It is used in high-pressure technology for the manufacture of glands, gaskets, various ampoules, insulators, etc.

Temperature Control

Creating the required temperature for high-pressure research is a multifaceted task and depends on the design of a particular apparatus. A simple case involves a possibility of external heating or cooling of a high-pressure device. The high temperature limit for the external heating is defined by the strength of the corresponding steels and alloys. This limit is about 500–700°C at pressures of several kilobars. Moderate temperatures of up to 300°C are achieved by immersing a high-pressure vessel in a thermostat with high-temperature polymethylsiloxane working fluid. In a similar thermostat, you can get moderately low temperatures of around −70°C using alcohols cooled with liquid nitrogen. The advantage of using thermostats and cryostats for temperature control in high-pressure enclosure is the uniformity of the temperature field, which determines the reliability of the results of temperature measurements and other physical quantities. However, in many cases, when it comes to pressures of the order of tens of thousands of kilobars, it is necessary to place a heater into the high-pressure enclosure. Here, the situation is highly dependent on pressure-transmitting medium. Any liquids, including polymethylsiloxane oil, decompose at high temperature on the heater coil with the release of soot, which limits obtaining temperatures above ∼500°C. This situation changes when using noble gases and nitrogen as a pressure-transmitting medium. In this case, it is possible to obtain temperatures in excess of 1,000°C. However, this raises another problem associated with convective heat transfer to the vessel walls. To obtain high temperatures in a device with a solid medium transmitting pressure, such as **Belt** or **Toroid**, short-circuited heaters (graphite or metal tubes) with resistance in fractions of ohms are used, which require power in hundreds of amperes at a voltage of a few volts. Temperature gradients in systems with similar heaters reach tens of degrees per mm. Heating of diamond anvils can be done externally upto several hundreds degrees. Much higher, but poorly controlled temperatures, are obtained using laser heating.

Temperature Measurement

The problem of temperature measurement in the high-pressure technique may be divided into three parts:

(a) Measurement of the temperature of the sample in the high-pressure enclosure, completely immersed in a thermostat or a cryostat. In this case, making sure there are no temperature gradients, you can use thermocouples, resistance thermometers and other sensors installed in the body of the enclosure outside the high-pressure zone.

(b) Measuring the temperature in an enclosure with an internal heater is a more complicated task. First, a temperature sensor should be inserted into the high-pressure device. As a rule, this sensor is a thermocouple, whose thermopower is weakly dependent on pressure. The thermocouple should be put in by the continuous way, without breaking and out of contact with other metals, in order to avoid side thermopower. For this purpose, in the case of hydrostatic enclosure, the electrical lead-throughs described in the **Electrical Leads** subsection can be used. In enclosures with a solid medium transmitting pressure, thermocouples are usually introduced through a narrow gap between the anvils, filled with medium. When a pressure is created, some outflow of material occurs, which often leads to rupture of thermocouples and other electrical inputs (see also **Electrical Leads**).

(c) For temperature measurement in diamond anvils at moderate temperatures of up to hundreds of degrees, created by miniature heaters, thermocouples attached to the anvils are used. Due to the high thermal conductivity of diamonds, the measured temperature closely corresponds to the temperature of the sample.

At temperatures created by a laser beam, the temperature is measured optically by the radiation of a heated body. Errors in temperature measurement in this case are very large.

Texture

Texture is the predominant orientation of the particles that make up the material, due to their specific shape, or formed as a result of directional effects. The formation of a certain texture of materials often occurs under uniaxial compression.

Toroid

A toroid is a high-pressure cell developed by Khvostantsev, Novikov and Vereshchagin in 1969[1] on the basis of a lentil cell (see **Bridgman Anvils, Lentils, Toroid**). The toroid cell is characterized by the presence of a torus surrounding the working space, which contributes to a more uniform distribution of stresses, increases the working volume, increases the limiting pressure and lifetime of the

---

[1]Khvostantsev L. G., Vereshchagin L. F. and Novikov A. P., *High Temp. High Press.*, 9, 637 (1977).

anvil, and finally allows entry of numerous electrical lead-throughs into the working cavity of the cell. Currently, the toroid is widely used for the synthesis of diamonds and new materials and in physical research. Further, the toroid is the only tool that allows one to study the diffraction of neutrons at pressures above 100 kbar. The large neutron centers in Grenoble, Los Alamos, Oak Ridge and ISIS (England) are equipped with toroid cells.

Tricritical Point

L. Landau noted that the line of phase transitions of the second order cannot simply end at some point, but it can go into the line of phase transitions of the first order. The point at which this occurs is described by Landau as a critical transition point of the second order. In modern literature, this point at the suggestion by Griffiths is called tricritical.[2]

---

[2]Stishov S. M., *Phase Transitions for Beginners*, Singapore: World Scientific (2018).

$\boxed{\text{V}}$

Valve

A valve is a device that allows control of the flow of liquids and gases. The simplest version of the valve used at high pressures is demonstrated in Fig. 81. The valve consists of a body 1, a stem 2, a base bushing 3 and a sealing gland 4, 5. The working end of the stem can be in the form of a cone 6 or sphere 7. The valve body is made of structural steel with a hardness of ~ 45–48 HRc. The stem is made of tool steel with a hardness of ~ 55–60 HRc. To work at pressures about 10 kbar, it is advisable to replace the mechanical drive with a hydraulic one.

Van der Waals

Johannes Diderik van der Waals (1837–1923) was a Dutch physicist and a Nobel Prize winner (1910). In his thesis, entitled "On the Continuity of the Gaseous and Liquid State of Matter", van der Waals developed a theory describing the liquid and gas phases of matter and the phase transition between them, including the critical point. The latter was discovered experimentally by the Irish physicist Andrews (see **Andrews**). The van der Waals equation, van der Waals loop, van der Waals forces — all have become firmly established in modern physics (for more, see any course in statistical physics).

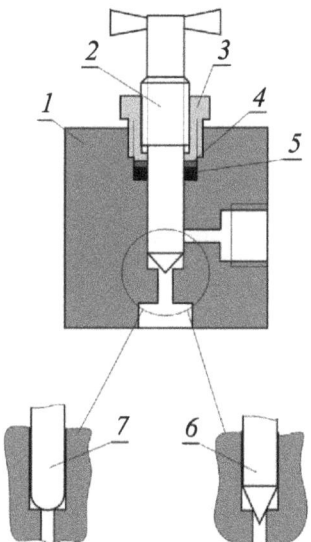

Fig. 81. The scheme of a simple high-pressure valve: 1 — body, 2 — stem, 3 — base bushing, 4, 5 — sealing gland, 6 and 7 — shapes of the stem.

Van Valkenburg Prize

The prize was established in honor of Alvin Van Valkenburg, one of the inventors of diamond anvils (see **Diamond Anvils**). Prizes are awarded to young scientists at the Gordon Conference on High Pressure (see **Gordon Conferences on High-Pressure**).

Viscosity

Viscosity is a property of liquid or gas that prevents fluidity. It is classified as dynamic viscosity η and kinematic viscosity $\nu = \eta/\rho$. In the context of the present book, dynamic viscosity is of interest. Most often, dynamic viscosity is measured in Poises (1Poise(P) = dyns/cm^2). It is useful to know a few numbers describing the viscosity of liquids used in high-pressure engineering:

Substance	Temperature	Viscosity (P)
Water	$20°\text{C}$	$1 \cdot 10^{-2}$
Glycerol	$20°\text{C}$	14.2
Liquid helium	$4\,\text{K}$	$3.3 \cdot 10^{-5}$
Ethyl alcohol	$20°\text{C}$	$1.1 \cdot 10^{-2}$
Light engine oil	$20°\text{C}$	1
Gasoline	$15°\text{C}$	$6.5 \cdot 10^{-3}$
Kerosene	$15°\text{C}$	$2.17 \cdot 10^{-2}$

Von Platen

Baltzar von Platen (Fig. 82) was a Swedish inventor, led by mystical ideas who,[1] built the first installation in the world capable of generating pressure and temperature sufficient to transform graphite into diamond. Von Platen built what is now called a split-sphere apparatus (Fig. 83). This apparatus is a thick-walled spherical vessel, cut into six identical parts (pistons). These parts, put together, form in the center of a cubic cavity, which is used to accommodate the tested materials. Pistons are isolated from each other by special materials, the main

Fig. 82. Baltzar von Platen (1898–1984).

purpose of which is to realize the possibility of moving them to compress a substance in a cubic cavity. After assembly, the sphere is covered with a rubber shell and placed in a large vessel in which hydrostatic pressure is created. As a result, the pressure in the center reaches large values. Essentially, the whole system is a three-dimensional implementation of Bridgman's anvils (see **Bridgman**

---

[1]Von Platen, A multiple piston, high pressure, high temperature apparatus, in *Modern Very High Pressure Techniques*, edited R. H. Wentorf, Jr. Washington: Butterworths (1962).

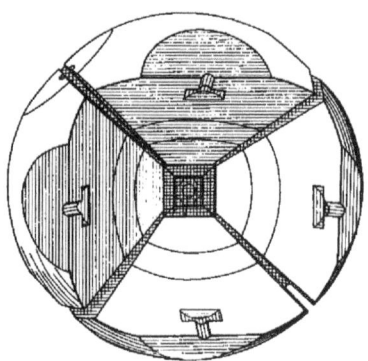

Fig. 83.   Von Platen's split sphere (von Platen, A multiple piston, high pressure, high temperature apparatus, in *Modern Very High Pressure Techniques*, edited R. H. Wentorf, Jr. Washington: Butterworths (1962)).

Anvils, Lentils, Toroid). This installation of von Platen served as a prototype for the subsequent development of multi-anvils devices.

It should also be mentioned that in the design of the low-pressure device of his installation, von Platen used the technique of winding steel strained tape,[2] which was subsequently widely used in the manufacture of presses and high-pressure vessels.

---

[2] *Ibid.*

$\boxed{\text{W}}$

Water

Water is a liquid substance consisting of H_2O molecules linked by hydrogen bonds. At a temperature of 0°C, water crystallizes with a decrease in density; therefore, as the pressure increases, the crystallization temperature decreases. However, at high pressure, crystalline water — ice — undergoes a phase transition, after which the melting point of ice begins to rise. As the pressure rises further, ice experiences a whole cascade of phase transitions (see Fig. 84). At a pressure of 2 GPa, ordinary hexagonal ice (Ice I) acquires a cubic structure (Ice VII). Then Ice VII at a much higher pressure (> 60 GPa) passes into the Ice X phase, which has lost its molecular structure and is an ionic crystal, where each hydrogen atom is symmetrically located between two oxygen atoms (Fig. 85). We note that an increase in the specific volume of water during freezing (crystallization) led to the creation of a specific high-pressure technique — the ice bomb method (see **Ice Bomb**).

White Dwarfs

White dwarfs are stars with masses of the order of the mass of the Sun, but with radii 100 times smaller. Accordingly, the average density of the substance of white dwarfs is 10^5–10^9 g/cm^3. White dwarfs are a product of the evolution of stars with a mass of less than 1.4 solar masses and result from the burning of hydrogen and helium as energy sources. The substance of white dwarfs is a system of fully

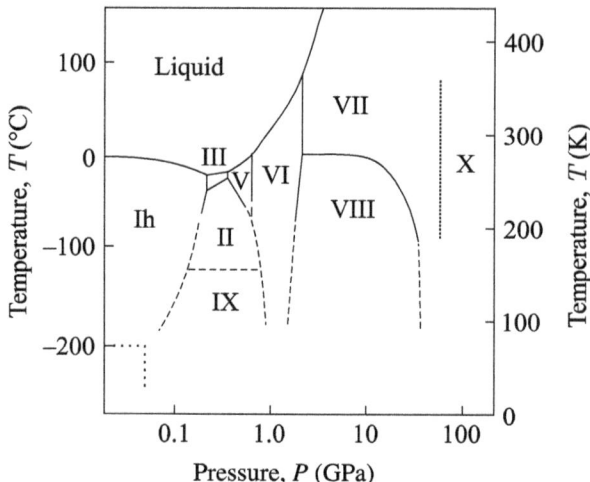

Fig. 84.   Phase diagram of water (Petrenko V. F. and Whitworth R. W., *Physics of Ice*, Oxford, UK: Oxford Univ. Press, (1999)).

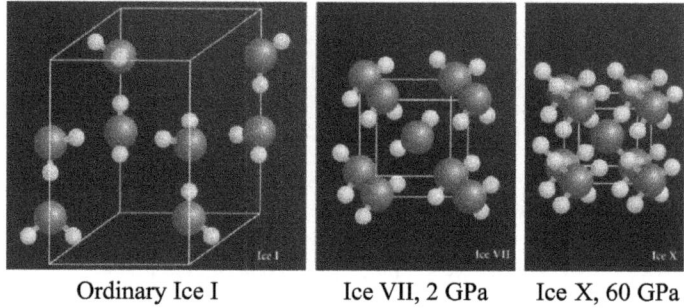

Ordinary Ice I   Ice VII, 2 GPa   Ice X, 60 GPa

Fig. 85.   The evolution of the crystal structure of ice at high pressures (Hemley R. J. and Ashcroft N.W., *Phys. Today*, 51, 8 (1998)).

ionized carbon and oxygen nuclei immersed in a sea of degenerate electron gas.[1] Such a system is able to crystallize with release of heat of crystallization, which, apparently, takes place in the white dwarfs, influencing their thermal history.[2] There is astronomical evidence

---

[1]Shapiro S. D., Teukolsky S. A. and Black Holes, *White Dwarfs and Neutron Stars*. New York: John Wiley and Sons (1983).
[2]Van Horn H. M., *Astrophys. J.*, 151, 227 (1968).

that most of the observed white dwarfs are in the process of crystallization, hence, given their composition, the ideas about diamonds shining in the Universe follow. However, it should be borne in mind that the nuclei of carbon and oxygen crystallize in white dwarfs and the product of crystallization apparently has little in common with diamonds.

Windows for Hard Radiation

Hard radiation generally describes electromagnetic frequencies from the extreme ultraviolet to X-rays and γ-rays and fast neutrons. Light elements that weakly absorb radiation (for example, lithium hydride LiH, epoxy resin: beryllium Be, boron B, carbon C (diamond), boron carbide B_4C) are used in windows for γ- and X-rays. In cases where a large aperture is not required (for example, when studying the Mossbauer effect or X-ray structural studies in the energy dispersive mode), beryllium and diamond windows can be installed as Poulter windows (see **Windows Optical**). However, in the case of a large aperture, the entire cell must be made of a material transparent to X-rays, e.g., beryllium or a diamond single crystal.[3] Pressed boron gaskets are used in systems with hard metal anvils. Diamond anvils with different gaskets are currently the priority tools for X-ray diffraction studies at high pressures.

For neutron scattering, cells made of materials with a small scattering cross-section, for example, aluminum alloys, or the same diamonds in the form of large anvils, are used. An alloy of titanium and zirconium (Ti–Zr), with zero coherent scattering, is used for the manufacture of high-pressure vessels for neutron research. The Ti–Zr alloy also serves as a gasket material in the toroid cell, adapted for neutron research. Cells made of sintered alumina also have been used in the neutron scattering experiments.

---

[3]Jamieson J. and Lawson A., Debye-Scherrer x-ray techniques for very high pressure studies, in *Modern Very High Pressure Techniques*, edited R. H. Wentorf, Jr. Washington: Butterworths (1962).

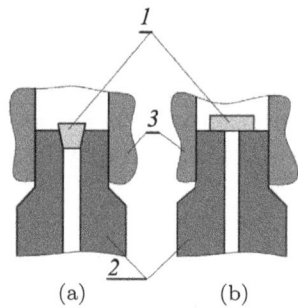

Fig. 86.   Optical windows for high pressures: (a) optical conical window, (b) flat optical window (Poulter window): 1 — windows, 2 — closure, 3 — high-pressure cell.

Windows Optical

Optical windows are used to conduct a variety of optical studies at high pressures. The materials of the windows (glass, quartz, sapphire, diamond, etc.) are selected for their optical and mechanical properties. Germanium and even salt can be used for special studies (see the original design of optical salt windows by H.G. Drickamer[4]). In Fig. 86, two popular window designs are shown.

In both cases, mating surfaces must be thoroughly processed to optical quality. The conical window before installation can be wrapped with lead or indium foil. However, the performance of the conical window is inferior to the flat window, which is called the "Poulter window". As a rule, the windows of quartz and sapphire are made so that the crystal axis C coincides with the axis of the window. Sapphire and glass windows with the right choice of parameters can work up to 20 kbar or more.

---

[4]Drickamer J. H. G. and Balchan A. S., High pressure optical and electrical measurements, in *Modern Very High Pressure Techniques*, edited R. H. Wentorf, Jr. Washington: Butterworths (1962).

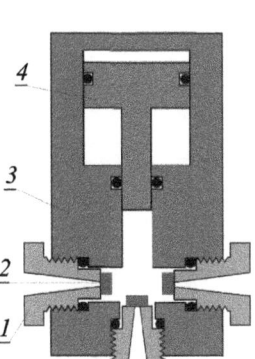

Fig. 87. Scheme of a high-pressure optical cell (Jamieson J. and Lawson A., in *Modern Very High-Pressure Techniques*, edited R. H. Wentorf, Jr. Butterworths (1962)): 1 — optical input, 2 — glass window, 3 — body, 4 — piston intensifier.

Figure 87 shows a scheme of a real optical cell used up to pressures of 25 kbar in the study of the Raman spectra of diamond.[5] As can be seen from the figure, the optical camera and intensifier, creating pressure, constitute a single unit, which, however, is not always convenient.

---

[5]Walley E. *et al.*, *Rev. Sci. Inst.*, 47, 845 (1976).

Conclusion

In conclusion, I would like to express the hope that this little book will help some scientists to become more comfortable in the field of high-pressure science and technology. I am greatly indebted to Dr. Marc Costantino for numerous remarks and corrections and Dr. Alla Petrova who helped with English versions of the figures. Finally, I am very grateful to all the scientists who encouraged me to publish the English version of *The ABCs of High-Pressure Science*.

Index

Lightning Source UK Ltd.
Milton Keynes UK
UKHW021814160421
382089UK00001B/114